U.S. Marine Corps Combat Water Survival

Plus

Marine Pistol Marksmanship

Department of Defense

Pentagon
Publishing

MCRP 3-02C

Marine Combat Water Survival

U.S. Marine Corps

PCN 144 000069 00

To Our Readers

Changes: Readers of this publication are encouraged to submit suggestions and changes that will improve it. Recommendations may be sent directly to Commanding General, Marine Corps Combat Development Command, Doctrine Division (C 42), 3300 Russell Road, Suite 318A, Quantico, VA 22134-5021 or by fax to 703-784-2917 (DSN 278-2917) or by E-mail to **morgann@mccdc.usmc.mil**. Recommendations should include the following information:

- Location of change
  - Publication number and title
  - Current page number
  - Paragraph number (if applicable)
  - Line number
  - Figure or table number (if applicable)
- Nature of change
  - Add, delete
  - Proposed new text, preferably double-spaced and typewritten
- Justification and/or source of change

Additional copies: A printed copy of this publication may be obtained from Marine Corps Logistics Base, Albany, GA 31704-5001, by following the instructions in MCBul 5600, *Marine Corps Doctrinal Publications Status.* An electronic copy may be obtained from the Doctrine Division, MCCDC, world wide web home page which is found at the following universal reference locator: **http://www.doctrine.usmc.mil**.

Unless otherwise stated, whenever the masculine gender is used, both men and women are included.

DEPARTMENT OF THE NAVY
Headquarters United States Marine Corps
Washington, D.C. 20380-1775

6 January 2003

FOREWORD

Marine Corps Reference Publication (MCRP) 3-02C, *Marine Combat Water Survival*, provides Marine Corps combat water survival techniques, procedures, and training standards. This publication also teaches Marines to cross water obstacles and perform water rescues correctly and safely.

This publication is the foundation for teaching Marines correct water survival techniques and procedures that are used throughout the Marine combat water survival program (MCWSP). Once an individual or a unit has completed the MCWSP, this publication can be used as a refresher course before water operations.

MCRP 3-02C supersedes Fleet Marine Force Reference Publication (FMFRP) 0-13, *Marine Combat Water Survival*, dated 16 September 1991.

Reviewed and approved this date.

BY DIRECTION OF THE COMMANDANT OF THE MARINE CORPS

EDWARD HANLON, JR.
Lieutenant General, U.S. Marine Corps
Commanding General
Marine Corps Combat Development Command

Publication Control Number: 144 000069 00

MCRP 3-02C Marine Combat Water Survival

Table of Contents

Chapter 1. Survival at Sea

Chapter 2. Water Rescues

Chapter 3. Treatment of Casualties and Avoidance of Dangerous Marine Life

Chapter 4. Negotiating Water Obstacles

Chapter 5. Fording Waterways

Appendix. Knot Tying

Throughout history, water has posed special challenges to Marines and Sailors during times of both peace and war. Therefore, the inherent dangers associated with waterborne operations demand that Marines and Sailors receive proper water survival training. Combat units that are confident in their ability to work in and around water can use the water to their advantage in combat, and history is filled with examples of how the proper preparation or training for survival in water reduced or averted disaster.

USS *Indianapolis*

On Sunday, 29 July 29 1945, the heavy cruiser USS *Indianapolis* (the *Indy*) was en route to the Philippine Sea. Shortly before midnight (about 39 hours out of port), the *Indy*, running blacked out and unescorted, was rocked by two explosions on her starboard side. With communications smashed, the ship could not signal its distress and sank within 15 minutes.

Three life rafts and a floater net supported a few survivors, but the rest of the survivors drifted about, held up by rubber life belts or Mae Wests. About 60 seamen died the first night. Survivors assumed the ship would be reported overdue in Leyte, and that they would be rescued within 2 days. By Monday evening, however, panic began to set in as some life jackets lost their buoyancy from the long immersion. Some men even fought over life jackets, which resulted in at least 25 deaths. No one dared sleep for fear of losing his jacket.

Throughout the next several days, in-transit aircraft flew nearby without spotting the desperate seamen. As best they could, the men kept together, some tying long ropes to each other, floating like corks on a net.

Not until late Thursday morning, 3 1/2 days after the ship sank, were the men discovered. Luckily, a plane on a routine flight over the area spotted the survivors. When ships picked them up that night, the survivors learned they had never been reported overdue. Every one of the 1,196-man crew was a casualty; 880 crewmembers were listed as dead or missing. Although many lives were lost, the innovative and expedient use of flotation devices and float techniques employed by the survivors helped save hundreds of lives.

USS *America*
On Thanksgiving eve, 23 November 1995, the USS *America* made its way through the Arabian Sea. Twenty-year-old, Marine Lance Corporal (LCpl) Zachary Mayo was unable to sleep and, wanting some fresh air, made his way onto an open-air platform near the aircraft hangar bay, which was three levels below the sleeping quarters. While he was on the platform, the ship veered suddenly, throwing LCpl Mayo through the platform's protective bars and into the sea, 30 feet below.

Frantic, LCpl Mayo called out in vain to the watchmen on the flight deck, which was 64 feet above him. It soon became clear to him that the USS *America* would keep its course into the Gulf of Oman until his absence was discovered at morning muster. The LCpl took a moment to consider his situation. Since land was at least 100 miles away, swimming was suicide; he would have to stay afloat until a search party found him.

Using the techniques he had learned during combat water survival training, LCpl Mayo made a flotation device out of his coveralls and tried to relax. Meanwhile, business continued as usual aboard the USS *America.* Since LCpl Mayo was on special assignment with the hazardous materials division, his absence went unnoticed until a petty officer asked several Sailors if they had seen their shipmate recently. By the time a roll call had been completed, LCpl Mayo had been adrift at sea for over 24 hours. Although three, fixed-wing Viking aircraft were deployed to search for LCpl Mayo, most people aboard ship feared the worst.

After 34 hours at sea, LCpl Mayo was discovered by fishermen on a Pakistani fishing boat. LCpl Mayo's survival is a testament not only to his incredible physical courage, but also to the soundness of the lifesaving training and techniques he received during combat water survival training.

Chapter 1
Survival at Sea

As a Marine, you face a variety of potential water emergencies whenever you cross expanses of water: ships, watercraft, and amphibious assault vehicles (AAVs) can sink; aircraft can crash into the sea; or you can accidentally fall into the water. However, there are some basic precautionary measures you can take to protect your safety and reduce your chances of becoming a water casualty. Determine the following information as soon as you board any type of vessel. Your knowing the following information may save your life or the life of your fellow shipmates.

- How many life preservers and lifeboats/rafts are on board?
- Where are the life preservers and lifeboats/rafts located?
- What type of unit survival equipment is on board?
- Are individual survival kits issued to each person on board?
- How much food, water, and medicine do the survival kits contain? When was the last time the contents were inspected for proper quantities and shelf life expiration?
- Is there sufficient survival equipment available for the number of personnel?
- How many other personnel are there on board, and where are they located?
- What are the egress procedures for the ship, boat, watercraft, AAV, or aircraft?

Abandoning Ship

When you embark on a Navy ship, you will receive abandoning ship instructions from Navy personnel. If given the order to abandon ship, report to your designated assembly area and put on a life preserver. DO NOT inflate the life preserver until you are clear of the ship. Torn life preservers will not inflate and inflated life preservers can block you, and those behind you, from exiting the ship. A flotation device that has been inflated may also burst if you jump from a significant height. See pages 1-15 through 1-35 for staying afloat with and without a life preserver.

DO NOT remove your clothing, boots, or shoes before abandoning ship. Your trousers and blouse may be the only flotation devices available if your life preserver is faulty or becomes damaged, and your clothes can provide some insulation from the cold water. However, remove your soft cover and place it in a cargo pocket for later use. The soft cover is both lightweight and a good protection against sunburn caused by the sun's rays reflecting off the water.

Jettisoning Equipment

Equipment should be kept properly packed and waterproofed in case you have to abandon ship. If entering the water from a height greater than 30 feet, wearing your equipment (e.g., pack, helmet, gas mask) could cause injury. Upon impact with the water, the helmet will "cup" air inside of it. The chin strap may also create a "hanging effect" as you submerge from the force of the fall. Therefore, you should remove your helmet and gas mask before abandoning ship.

If you are unable to maintain buoyancy due to the amount of equipment secured to your pack and body, then jettisoning some of your equipment may become necessary. Equipment that you should always retain includes canteens of freshwater, first aid kit, soft cover, and survival kit. The survival kit should include first aid items, water purification tablets or drops, fire starting equipment, signalling items (e.g., flashlight, strobe light, chemlights), food procurement items, and shelter items. Other items in a survival kit include sunburn lotion and lip balm, knife, goggles/sunglasses, plastic bag, matches and lighter, mirror. See Marine Corps Reference Publication (MCRP) 3-02F, *Survival,* for a detailed list of survival items and applications.

Abandoning Ship Technique

When abandoning ship, safety considerations must be observed. Use the following technique when abandoning ship without your combat gear:

Place your hands on their opposite shoulders, forming a criss-cross pattern.

Step to the edge of the ship's deck and check the water below for debris or survivors. If the water is clear, look straight ahead and prepare to jump. If the water is not clear, move to another location.

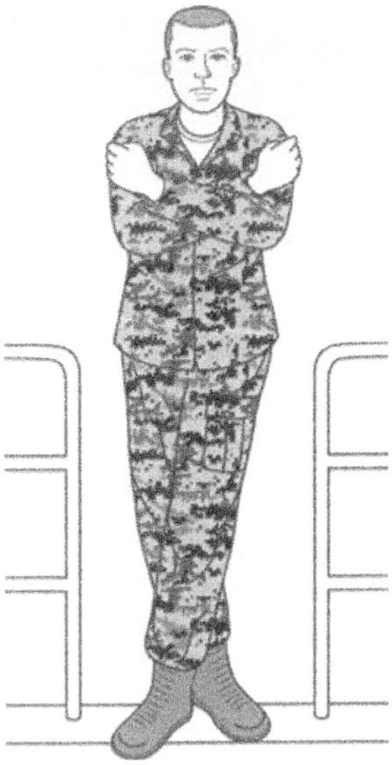

NOTE: DO NOT hold your nose as you abandon ship. If you do hold your nose, the force of impact into the water could jar your arm and hand and cause you to break your nose.

Step off the side of the ship with a smooth, 30-inch stride. DO NOT DIVE OFF THE SHIP. DO NOT LOOK DOWN AT THE WATER. LOOK STRAIGHT AHEAD. Looking down at the water can render you unconscious or cause injuries upon impact.

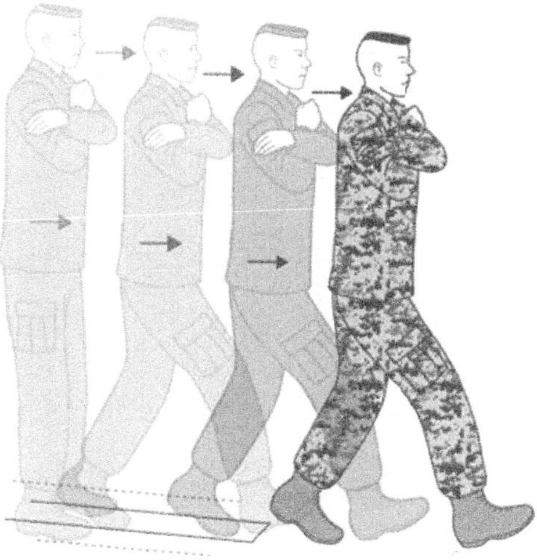

Bring your trailing leg forward during the fall. Cross your trailing leg behind your leading leg.

Keep your head parallel to the water's surface until hitting the water.

You should remain in the abandon ship position until your descent into the water has almost stopped. However, the weight imbalances in your body may cause you to be in a "J" shape under the water

Once your downward motion has ceased, your feet may be parallel with the ocean bottom or you may be nearly inverted with your feet over your head. To counteract potential disorientation, you should pause briefly and allow the natural buoyancy of your torso to bring your body to a nearly upright position.

Floating debris can cause hazards. Therefore, you should swim upward, extending one arm (hand is shaped as a fist) upward to feel for obstructions. If you encounter debris, try to push it away or surface in a different location.

Swim away from the ship. DO NOT LOOK BACK AT THE SHIP. Looking back could slow your movement away from the area. Remember, your objective is to leave the area as quickly as possible because—

- Equipment and debris may be falling from or spilling out of the ship.
- Additional casualties can occur if individuals abandoning the ship fall on top of swimmers already in the water.
- Swimmers close to the sinking ship may get pulled underneath the water by the suctioning effect of the ship as it goes under.

Modified Abandoning Ship Technique

When abandoning ship while wearing full combat gear (weapon, helmet, and a properly waterproofed pack), safety considerations must be observed. The modified abandoning ship technique is used while wearing full combat gear and exiting from a height less than 30 feet.

WARNING

When exiting from a height greater than 30 feet, remove your helmet and pack. Fasten the helmet to the pack or place it inside the pack before jettisoning. If jettisoning gear from the ship or aircraft, check the water below for survivors before throwing the gear forward of the intended jump area. Once in the water, retrieve your gear and swim out of the area.

Place your weapon over one shoulder, muzzle down, with the weapon parallel to your side.

Place your arm and hand along the weapon and hold it to your side.

Take your free hand and place it on top of your helmet to prevent neck/spinal injury from the force of the water pulling upward on the helmet as your body enters the water.

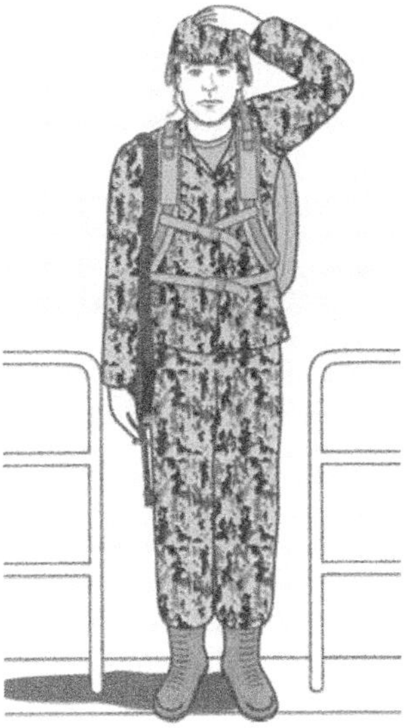

NOTE: An alternate method is to place your weapon over one shoulder, muzzle down, with the weapon parallel to your side. Reach across your body and grasp the sling of your weapon and hold it to your body. Take your free hand and place it on top of your helmet.

Step to the edge of the platform and check the water below for debris and survivors. If the water is clear, look straight ahead and prepare to jump. If the water is not clear, move to another location.

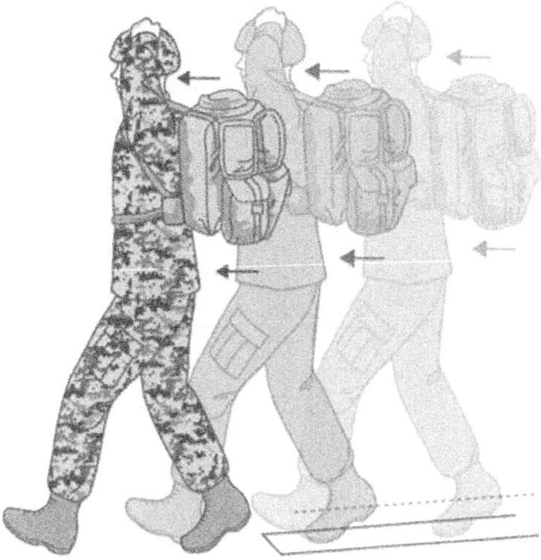

Step off the side of the platform with a smooth, 30-inch stride. DO NOT DIVE OFF THE SHIP, DO NOT LOOK DOWN AT THE WATER, LOOK STRAIGHT AHEAD. Looking down at the water can render you unconscious or cause injury upon impact.

Bring your trailing leg forward during the fall. Cross your trailing leg behind your leading leg.

Keep your head parallel to the water's surface until hitting the water.

You should remain in the modified abandoning ship position until your descent into the water has almost stopped. The buoyancy of a properly waterproofed pack will immediately pull you to the surface. Once you break the water's surface, unsling your weapon and loop the sling over your head. You may have to seesaw the sling so as to ensure the sling passes between your helmet and pack. Once over your head, loop the sling around your neck with the weapon aligned in the center of your body, muzzle down. Lean back on your pack and perform the combat travel stroke to exit the area.

> *Note:* You may also remove the pack (unsnap the quick-release strap on one side of the pack, and pass the strap to your other hand in order to maintain contact with the pack, which is functioning as your flotation device). Remain horizontal in the water with your pack under your chest. Perform a breast stroke kick.

Surface Burning Oil Swim

After you have abandoned ship, rise to the surface using the techniques shown on the page 1-6. However, you must remember that fuel from sinking ships or downed aircraft will float on the surface of the water. Therefore, you must move clear of the floating fuel by swimming away from the ship or aircraft as soon as possible. Either swim upwind (into the wind) of the ship/aircraft or swim against the current. Either method allows you to move away from the fuel and the wind/current will push the fuel past you. To properly execute a surface burning oil swim—

Extend your arms overhead as far as possible.

Wave your arms back and forth vigorously to splash a hole while moving upward.

Splash as long as possible to push burning fuel away from the surfacing area.

Use your arms and hands to sweep away fuel and debris.

Kick your legs in a constant breast stroke kick.

Extend your arms (palms outward) forward on the surface, arms shoulder-width apart.

Pull your hands in and back toward the chest.

Stop your hands in front of your face and rotate them so that your palms face forward (roughly halfway out of the water).

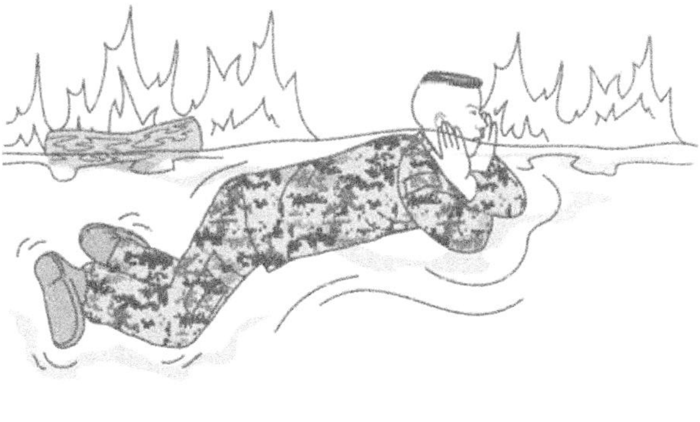

Sweep your arms forward to a full extension at the shoulder width. This splashes debris, oil, or burning liquids aside. To reduce the chance of fatigue, use two short splashes to the front to extend the path.

Repeat the preceding step as necessary while swimming clear of the area.

Surviving With a Pack

If packed properly, your pack will float, and it is your key piece of equipment for staying afloat and overcoming water obstacles. If the pack's contents are properly waterproofed, it can support you (with a combat load) in the water. Buoyed up by a waterproofed pack, you eventually emerge from the water with all your equipment (e.g., boots, helmet, flak jacket, weapon, survival items).

Your pack floats based on a scientific principle known as Archimedes' principle. This principle states that an object submerged in a liquid is buoyed up by a force equal to the weight of the liquid displaced (pushed aside) by the object. If the weight of the displaced liquid is greater than the weight of the object, the object floats. If the weight of the displaced liquid is less than the weight of the object, the object sinks. For example, a machine gun sinks in the water, but it still weighs less in the water than it does on land. Even though a machine gun sinks, it is still buoyed up by a force equal to the weight of the water it displaces. For this reason, you should not try to hold yourself or your equipment any higher out of the water than they would naturally float; doing so wastes both energy and body heat.

Preparing Equipment

Before packing your pack, you must prepare your gear/equipment. Tape or pad all sharp edges and equipment corners. Ideally, your gear/equipment is placed in plastic bags and the plastic bag is then placed inside the standard-issued, rubberized, waterproof bag. Waterproof bags are not completely water tight; henceforth, the added protection of first wrapping the gear/equipment in a plastic bag, then placing the plastic bag inside the waterproof bag. Large plastic bags (e.g., trash bags) work well for bulky equipment (e.g., sleeping bags, field jackets, shelter halves, gas masks). Small plastic bags work best for small items (e.g., shaving gear).

> *Note:* If a gas mask must be carried outside the pack, cover it with a waterproof bag.

You will need the following items, which are available through the supply system, to prepare your equipment for packing:

- Waterproof bags:
  Pistol bag, plastic size #1 (8" x 18").
  Rifle bag, plastic size #2 (10" x 56").
  Machine gun bag, plastic size #3 (15" x 56").
  Multipurpose bag, cover, plastic size #4 (20" x 84").

- Panama work vest (life preserver).
- Small plastic bag, self sealing (6" x 6" and 12" x 12").
- Riggers tape, olive drab (2 1/2").
- Willie peter bag, waterproof clothing.

Tying Waterproof/Plastic Bags

Try to remove the excess air from the waterproof/plastic bag before securing the bag's opening. If filled with air, the bag can burst if pressed from the outside. Perform the following steps to tie a waterproof/plastic bag:

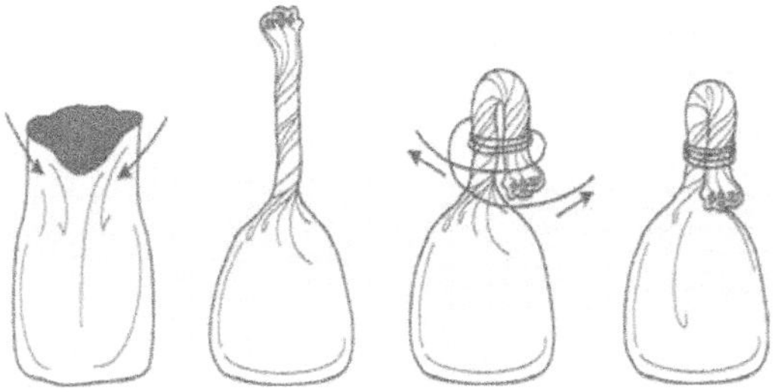

Packing the Pack

Place filled and tied bags inside the pack. Carefully handle sharp items (e.g., tent pegs, poles) to prevent puncturing the bag. Place items in the pack in order of expected use. Close the pack and its compartments. Attach sleeping mats or bags as high as possible on the outside of the pack.

Swimming With the Pack

A useful technique for propelling yourself forward while wearing a pack is the combat travel stroke. This technique allows you to float nearly horizontal and to propel yourself forward with bicycle-style kicks and breast stroke-style arm sweeps. To execute the combat travel stroke—

Body position
Keep the upper part of your body prone to the water, let your legs dangle horizontally, and keep your face up.

Arm action
Extend your hands out in front of your waist. Sweep your arms slightly downward and back 90 degrees to propel your body through the water. Move your hands back to the front of your waist. Repeat.

Leg action
Continuously move your legs in a bicycle-like movement, bringing your knees up high and step out.

Breathing
Keep your face out of the water during the stroke, and breathe freely.

Staying Afloat With a Life Preserver

The best form of flotation is to find any kind of floating object that will keep you and your equipment out of the water or minimize your exposure to the water. Life preservers are the best method, as they allow you to wear your clothes for heat retention and sunburn prevention. Marines use two basic classes of life preservers: inherently buoyant life preservers and inflatable life preservers.

Inherently Buoyant Life Preservers

Inherently buoyant life preservers are either vest-type (worn like a jacket) or yoke-type (worn around the neck). The preserver's outer envelope is either a cotton or water resistant material that encloses a removable fibrous glass or plastic foam filling.

In the Marine Corps, the most common type of inherently buoyant life preserver is the vest-type with collar, known as the kapock preserver. The kapock consists of collar straps, upper front chest straps, leg straps, and waist drawstrings that secure the preserver to you. The leg straps, which are fitted on both sides of the life preserver, ensure that the preserver remains around your chest while you are in the water. A chest strap is attached to the life preserver to facilitate lifting you out of the water. The strap can also be attached to other survivors or to lifeboats to reduce the fatigue that results from holding onto a floating/secured object by hand.

Inflatable Life Preservers

Marine Corps aircraft and AAVs have inflatable life preservers on board. The inflatable life preserver issued to Marines is known as the LPP (life preserver personal). The LPP is capable of both oral inflation and CO_2 cartridge inflation. The LPP consists of buoyancy chambers, CO_2 inflator, and an oral inflation tube. The buoyancy chamber is deck gray in color and is made from a neoprene-coated nylon fabric.

Inflatable life preservers must be stored in a cool, dry place. Heat, moisture, and light cause deterioration of the life preserver material. Do not stow CO_2 cylinders near steam lines or radiators. Heat can increase the pressure inside the cylinders causing them to explode. Avoid sharp edges in stowage. Sharp edges increase wear and tear on the life preservers and may also puncture inflatable buoyancy chambers.

> **CAUTION:** Do not inflate the life preserver until you are clear of the aircraft, ship, or vehicle. Torn life preservers will not inflate and inflated life preservers can block you, and those behind you, from exiting the aircraft, ship, or vehicle.

To don and adjust the LPP—

Remove the LPP from its storage container.

Fasten the belt fasteners in front with the pouch in the rear.

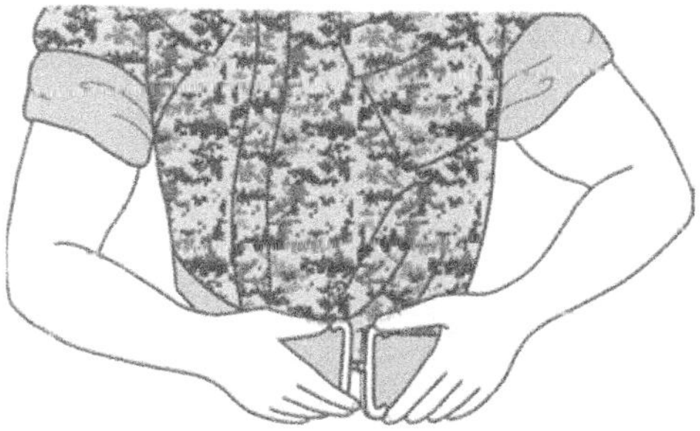

Adjust the belt to fit; secure any excess belt by mating the hook and pile tape.

Rotate the pouch to the front and re-adjust the belt if necessary.

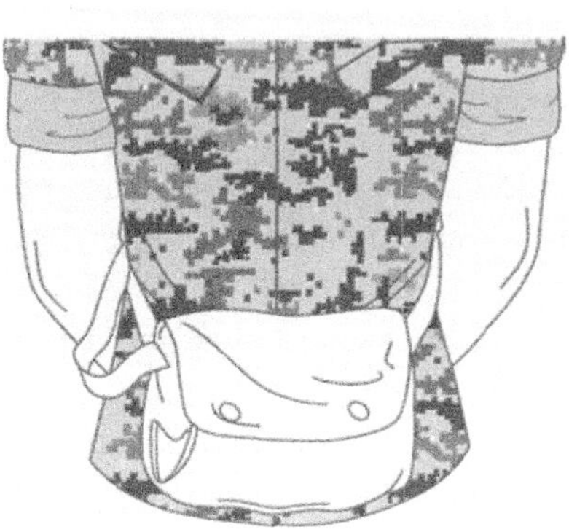

Open the snap fasteners on the pouch and unfold the life preserver.

Place the deflated life preserver over your head.

Place the storage container into the pouch after donning the life preserver.

Lift the lower end of the life preserver out of the pouch.

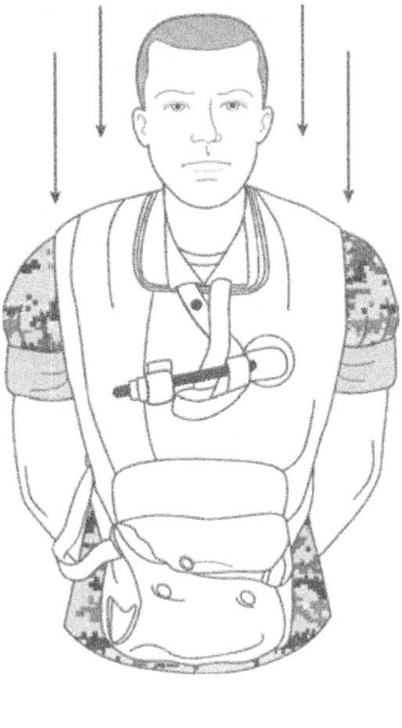

Inflate the life preserver by pulling on the lanyard attached to the CO_2 inflation valve or by blowing on the end of the oral inflation valve.

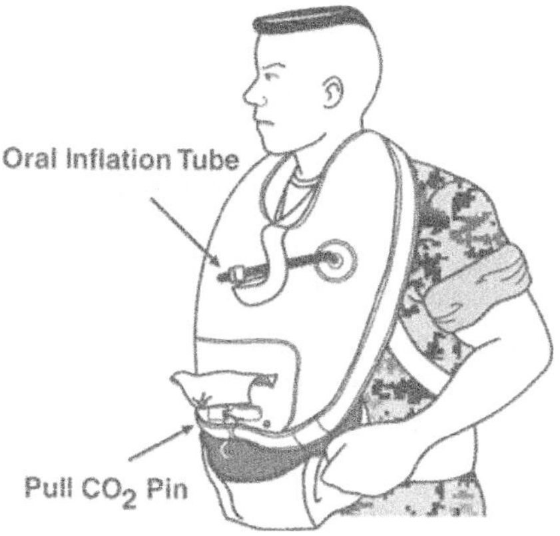

Staying Afloat Without a Life Preserver

You may be in open water without any floating objects or a life preserver to help you survive. If so, uniform blouses and trousers can be made into expedient flotation devices.

Floating With an Inflated Blouse

It is possible to float by a bubble of air trapped in the shoulders of your blouse. The air rises to the back and shoulders of the blouse and supports you at the water's surface. An inflated blouse is also a temporary flotation device used by weaker swimmers while trying to remove their trousers. There is a primary and an alternate way to create a bubble of trapped air in a blouse—

Primary Method
Turn the collar inside the blouse to help create a seal.

Unbutton top button and pull collar around mouth and nose.

Take a deep breath and bend forward slightly at the waist. Exhale one-half to three-quarters of a breath into the blouse.

Grasp and twist the collar with one hand to create a seal, this prevents air from escaping out from the collar.

Use your free hand and feet to stroke and kick to the surface.

Gather and hold the blouse tightly at the collar and stomach level to prevent the blouse from losing air if it floats up too high.

Splash water on the blouse periodically to prevent the material from drying, dry material allows air to escape.

Repeat inflation as required.

Alternate Method
Turn the collar inside the blouse to help create a seal.

Unbutton the second button from the top.

Take a deep breath and bend forward slightly at the waist.

Place your mouth and nose inside the hole created by the open button and exhale one-half to three-quarters of a breath into the blouse.

Grasp material at the unbutton portion and pull downward.

Use your free hand and feet to stroke and kick to the surface.

Splash water on the blouse periodically to prevent the material from drying, dry material allows air to escape.

Repeat inflation as required.

Floating With Inflated Trousers

In warm water, trousers can be used as a primary expedient flotation device. However, in cold water, submerging your head to remove and inflate your trousers results in heat and energy losses that negate the benefit of using the trousers as a flotation device. Once your trousers are inflated, you float motionlessly as if wearing a life preserver. If needed, assume the heat escape lessening posture (known as the HELP position, see page 1-37) to slow heat loss. As trousers dry, air leaks out of the legs. To slow this process, occasionally splash water on the fabric. Reinflate trousers as needed.

Sling Method

The sling method works if you are a strong swimmer or naturally very buoyant. Take the following steps to inflate trousers using the sling method:

Take a deep breath, bend over, and remove your boots.

NOTE: Retain your boots. Tie the boot laces together and suspend the boots from your blouse or hang them around your neck so that they rest on your chest.

Remove your trousers. Button or zip the trouser's fly closed. This allows you to control airflow.

Tie the bottoms of the trouser legs in a square knot. (The appendix illustrates various knots.)

Ensure that the front (fly) of the trousers faces you.

Hold the trousers above the water's surface and behind your head. Grasp both sides of the waistband and open with both hands.

Kick strongly to stay on top of the water while slinging the trousers overhead in order to trap air into them.

Once the waistband is submerged in the water, air is trapped in the legs.

Hold and seal the waistband underwater.

Slip the inflated legs over your head. Hold the waistband in toward your chest, the fly facing your body. To prevent air from escaping from the trousers, seal the waistband by either folding it or twisting it.

Lie back and relax, resting the back of your neck against the knot.

Splash water on the trousers periodically to prevent the material from drying. Dry material allows air to escape.

To replenish air in the trousers, you will use a technique known as the scooping method. With one hand on the open waistband, extend the trousers in front of you just below the surface of the water and scoop air bubbles with your free hand into the open waistband until the trousers have sufficient air. Repeat as necessary.

Splash Method

The splash method is an alternative to the sling method. As with the sling method, you must kick strongly to remain at the surface. To inflate trousers using the splash method, perform the following:

Take a deep breath, bend over, and remove your boots.

NOTE: Retain your boots. Tie the boot laces together and suspend the boots from your blouse or hang them around your neck so that they rest on your chest.

Remove your trousers. Button or zip the trouser fly closed. This allows you to control airflow.

Tie the bottoms of the trouser legs in a square knot. (The appendix illustrates various knots.)

Ensure that the front (fly) of the trousers faces you.

Hold the trousers at the water's surface out in front of you by the waistband with the fly up.

Grasp the waistband at the surface with one hand. Insert your free hand into the waistband, palm down.

Flutter your hand rapidly to create bubbles. This sends a mixture of water and air bubbles into the trousers. The water passes through the fabric. The air remains trapped in the legs.

Hold and seal the waistband underwater.

Slip the inflated legs over your head. Hold the waistband in toward your chest, the fly facing your body. To prevent air from escaping from the trousers, seal the waistband by either folding it or twisting it.

Lie back and relax, resting the back of your neck against the knot.

Splash water on the trousers periodically to prevent the material from drying. Dry material allows air to escape.

To replenish air in the trousers, you will use a technique known as the scooping method. With one hand on the open waistband, extend the trousers in front of you just below the surface of the water and scoop air bubbles with your free hand into the open waistband until the trousers have sufficient air. Repeat as necessary.

Blow Method

The blow method is an alternative to the sling method. Use the blow method if you are a weak swimmer. Take the following steps to inflate trousers using the blow method:

Take a deep breath, bend over, and remove your boots.

NOTE: Retain your boots. Tie the boot laces together and suspend the boots from your blouse or hang them around your neck so that they rest on your chest.

Remove your trousers. Button or zip the trouser fly closed. This allows you to control airflow.

Tie the bottoms of the trouser legs in a square knot. (The appendix illustrates various knots.)

Ensure that the front (fly) of the trousers faces you.

Hold the trousers at the water's surface. Grasp both sides of the waistband and open with both hands.

Take a deep breath.

Drop 2 feet below the water's surface, pulling the waistband underwater.

Hold the waistband open with both hands and blow air into the trousers.

To fill the trousers with air, surface while keeping the waistband underwater, breathe in again, drop below the water's surface, and blow air into the trousers. Repeat these steps until the trousers are filled sufficiently. Once trousers are filled—

Hold the waistband underwater. Twist and pinch it off.

Slip the inflated legs over your head. Hold the waistband in toward your chest, the fly facing your body. To prevent air from escaping from the trousers, seal the waistband by either folding it or twisting it.

Lie back and relax, resting the back of your neck against the knot.

Splash water on the trousers periodically, to prevent the material from drying. Dry material allows air to escape.

To replenish air in the trousers, you will use a technique referred to as the scooping method. With one hand on the open waistband, extend the trousers in front of you just below the surface of the water and scoop air bubbles with your free hand into the open waistband until the trousers have sufficient air. Repeat as necessary.

Avoiding Heat Loss in Cold Water

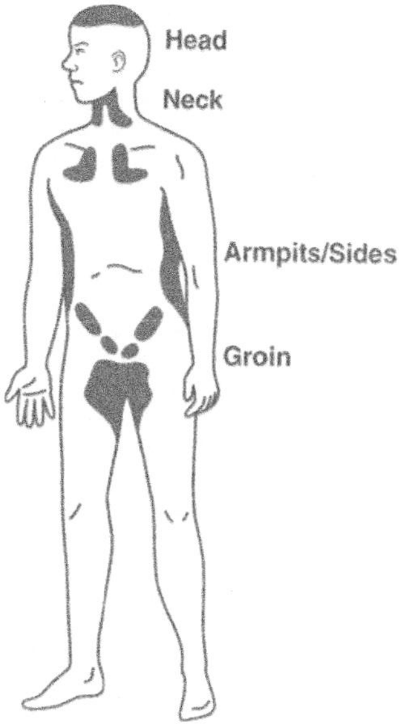

The rate of heat exchange in the water is about 25 times greater than it is in air of the same temperature. When you are immersed in cold water, hypothermia occurs rapidly due to the decreased insulating quality of wet clothing and as a result of water displacing the layer of still air that normally surrounds the body. You also lose about 50 percent of your body heat through your head; therefore, keep your head out of the water. Other areas of high heat loss are the neck, the armpits/sides, and the groin.

In cold water, DO NOT SWIM TO STAY WARM. Swimming, even with a slow and steady stroke, produces a lot of heat that is lost in the water. The heat loss can produce hypothermia that slows body functions and can result in serious injury or death. Remaining motionless conserves body heat three times longer than swimming. SWIM only if you have a flotation device and the shoreline is visible.

Individual Protection From the Cold

If you are equipped with a life preserver, assume the heat escape lessening posture (known as the HELP position) to slow heat loss and to protect major blood vessels near the body's surface. These areas lack insulating fat and are vulnerable to the chilling effects of cold water. To assume the HELP position—

Tuck your chin down tightly to cover your throat.

Draw your legs up in a fetal position to protect the groin.

Place your arms across your chest, tuck your hands into your armpits.

Wear some type of head covering (e.g., stowed cover, towel, handkerchief) to lessen heat loss through the scalp if head covering is available.

Group Protection From the Cold

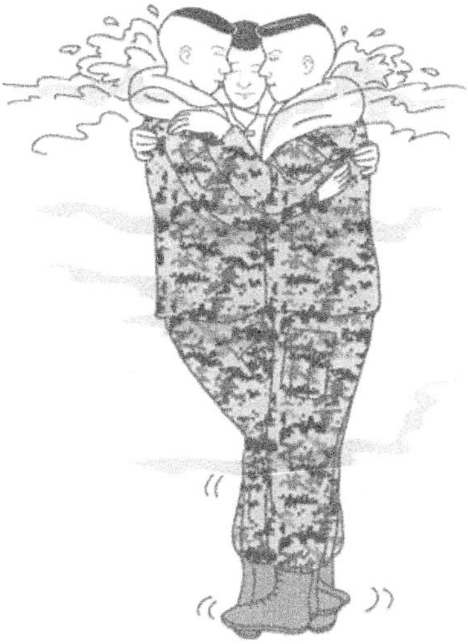

If three or more Marines are in the water and are equipped with life preservers, they should wedge tightly together and lock arms to form a circle known as a huddle position. This position protects vulnerable areas from heat loss.

A casualty who is suffering from the effects of the cold can be placed within the huddle to be surrounded by warmer water. If in the water for a prolonged period, it is recommended that everyone be rotated inside the huddle to maintain or re-warm each person's internal core temperature. If there are more than five Marines, they should make clusters of huddle positions.

Contact with other swimmers provides survival advantages—

Creates a larger target for search and rescue aircraft.

Provides additional warmth in cold water.

Improves morale.

Re-establishes the chain of command.

Reduces shock and panic.

Provides opportunities to administer first aid.

Supports exhausted Marines.

Drownproofing Methods

An object that floats on the surface has positive buoyancy. An object that floats a few feet beneath the surface has neutral buoyancy. An object that sinks has negative buoyancy. Most people have positive buoyancy and will float at the water's surface. To test your buoyancy—

Stand in water that is at your chest level or deeper.

Take a full breath.

Bend forward slowly.

Relax and wait.

If you have positive buoyancy, you will slowly rise to the surface. If you have neutral buoyancy, you will float a few feet beneath the surface. If you have negative buoyancy, you will sink. Regardless of whether you naturally float or sink, you can remain at the surface for extended periods without a life preserver if you exercise the appropriate drownproofing methods or survival strokes, which are based on your buoyancy.

Drownproofing methods consist of the T-method and the sweep. The breast stroke, side stroke, and elementary backstroke are the most common survival strokes. With any drownproofing method or survival stroke, remember the acronym **SAFE**:

SLOW EASY MOVEMENTS
Move slowly to conserve energy and minimize heat loss.

APPLY NATURAL BUOYANCY
Use natural buoyancy to support the body.

FULL LUNG INFLATION
Fill the lungs with each breath. Do not hold air in the cheeks.

EXTREME RELAXATION
Tight muscles are denser than relaxed ones and do not float as well.

Crawl Stroke

The crawl stroke, sometimes called the front crawl or free style, is the fastest stroke. To execute the crawl stroke—

Body position
Lie horizontal, on your stomach, in the water.

Look forward and downward at a 45 degree angle with the waterline between your eyebrows and hairline. Your head position is important as it assists in cutting a path through the

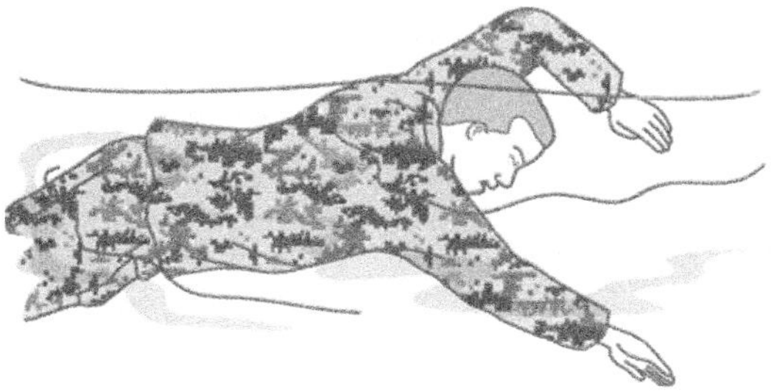

water. If your head is too high in the water, your lower body sinks significantly, making your stroke less efficient. If your head is too low in the water, water washes over your shoulders and neck, causing unnecessary drag.

Arm action

Arm action occurs in three phases: catch, propulsion, and recovery.

Fully extend one arm forward of your body, this positions your hand to catch the water in preparation for the propulsion phase.

To catch the water, bend your wrist (with your palm pointing outboard) and make an "S" shape (or inverted "S" shape) with your hand, ensuring that your hand does not cross the center of your body. Your left hand makes an "S" shape and your right hand makes an inverted "S" shape with your hand finishing at shoulder level.

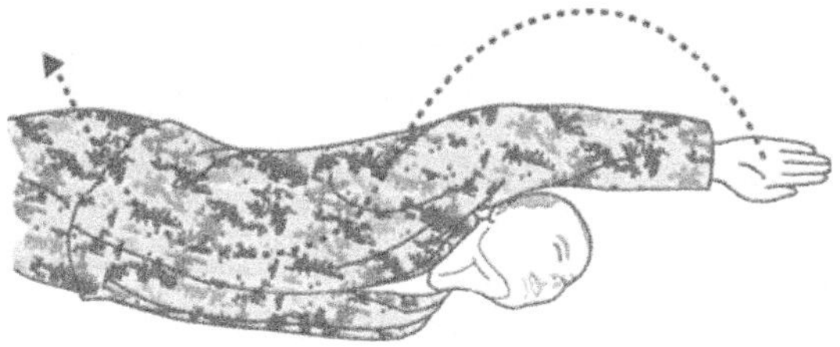

Push with your hands in a rearward fashion toward your feet until your arm is fully extended along your side, keeping your hands close to your body.

To begin the recovery phase, bend your arm at the elbow and raise your hand out of the water. Your hand breaks the surface of the water and maintains a height of 2 to 3 inches above the water's surface.

With your hand and arm moving in a forward manner, bring your hand past your head until your arm is about three-quarters of the way extended. At this point, turn your hand with your palm outboard so as to allow your thumb and forefinger to enter the water first.

Once your hand enters the water, continue to push your arm forward until it is fully extended.

You are now prepared to catch the water again. These steps are performed in an alternating pattern: when one arm is catching and propelling, the other arm is recovering.

Leg action

Use a flutter kick to create the leg action for the crawl stroke. This kick is used for both propulsion and keeping the lower body horizontal with the water's surface. The flutter kick is an alternating leg action: one leg is kicking in a downward motion while the other leg is recovering to the surface to prepare for the next kick. The size of the kick ranges from 12 to 15 inches and depends on your height.

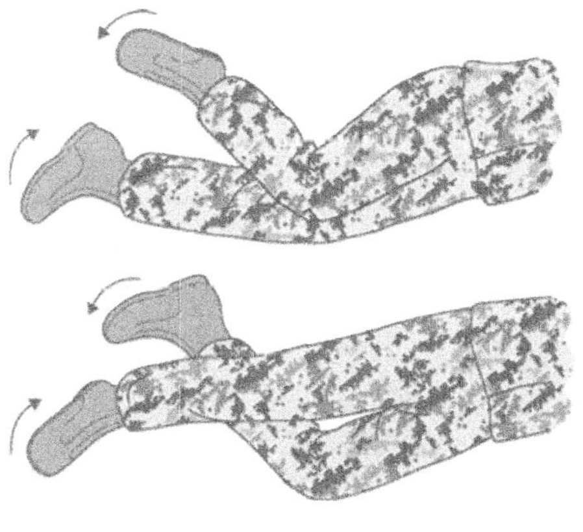

Maintain your legs in a semi-rigid manner.

Generate power for the kick from the hips.

Keep your feet loose at the ankles, they trail behind your legs and act as "flippers."

Execute the propulsion phase of this kick with a downward thrust of your leg.

Execute the recovery phase by pushing your leg back to the surface. This phase is complete when your foot reaches the surface.

Breathing
Breathe during the recovery phase of the arm action.

Roll your body and rotate your head to the side of your body where the arm recovery is occurring, this rolls the water away from your mouth. Keep your chin pushed back toward your shoulder.

Exhale while your face is still submerged and inhale when your face breaks the surface.

Breathe either bilaterally or rhythmically. To breathe rhythmically, breathe on the same side of your body every time your arm cycle occurs. Bilateral breathing requires you to breathe every one and a half arm cycles. To breathe bilaterally, breathe when your right arm is recovering, your face goes back in the water and the next time you breathe is when your left arm is starting its recovery three arm strokes later.

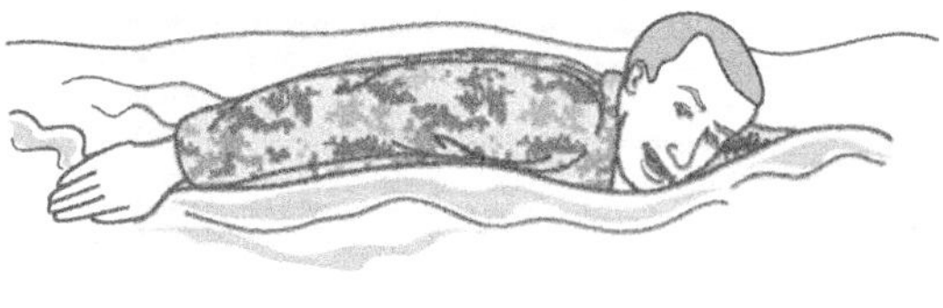

Bilateral breathing is the preferred method for breathing; it prevents the chance of hyperventilation and allows your body to maintain a lateral position in the direction you are swimming.

Coordination

This stroke uses constant arm and leg action.

T-Method

The T-method is a basic drownproofing method. This is the best survival technique if you have negative buoyancy. To execute the T-method—

With your face out of the water, take a deep breath and submerge your face in the water while holding your breath.

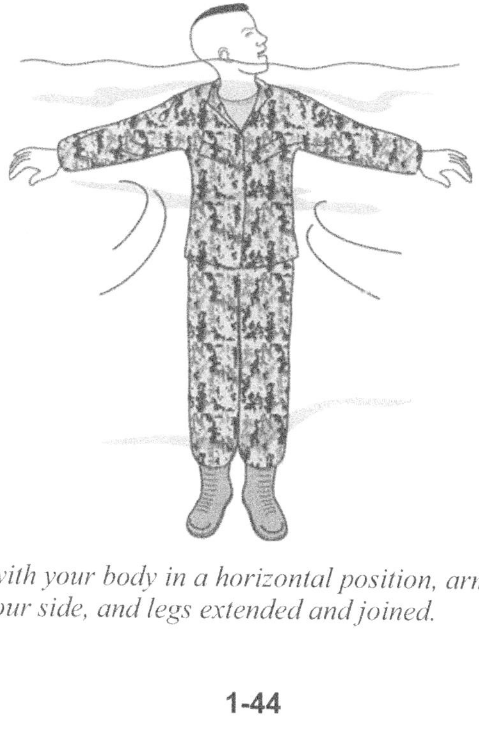

Float with your body in a horizontal position, arms extended from your side, and legs extended and joined.

Move your hands up to your armpits by tracing an imaginary line along your ribs.

Extend your arms outward (horizontal) to your sides, your body position resembles the letter "T."

Step out and forward with one leg and point your other leg to the rear, your knees should be slightly bent. Simultaneously, bring your arms down to your sides. Then bring your legs back together. You complete the steps by exhaling most of your air and preparing to surface your face to obtain another breath of air. Hold your head out of the water. Tilt your head back slightly. Breathe normally.

Once your breath is complete, move your hands up and down directly in front of your body. Do this two or three times to slow your descent into the water.

> NOTE: To avoid hyperventilating, hold your breath below the surface of the water for no more than 10 seconds.

The Sweep

The sweep works well if you have slight to excellent positive buoyancy. To execute the sweep—

Float face down in the water: bend 45 degrees at the waist, arms and legs dangling, head hanging down, relax all muscles.

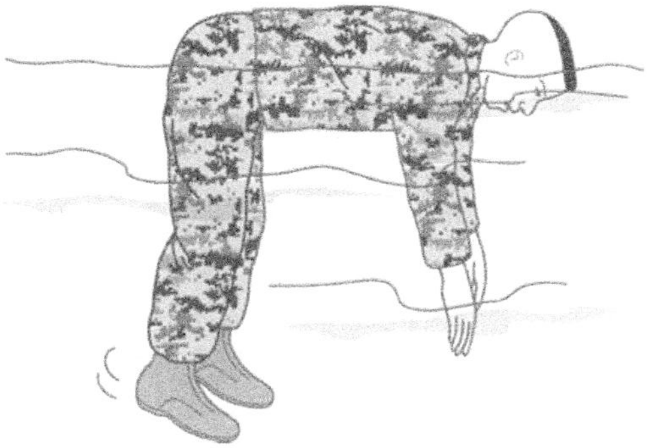

Spread your feet slowly to prepare for a single kick (one leg is forward and one leg is rearward).

Cross your arms in front of your chest, palms outboard with the back of each hand touching the opposite ear.

Exhale prior to raising your head for a breath.

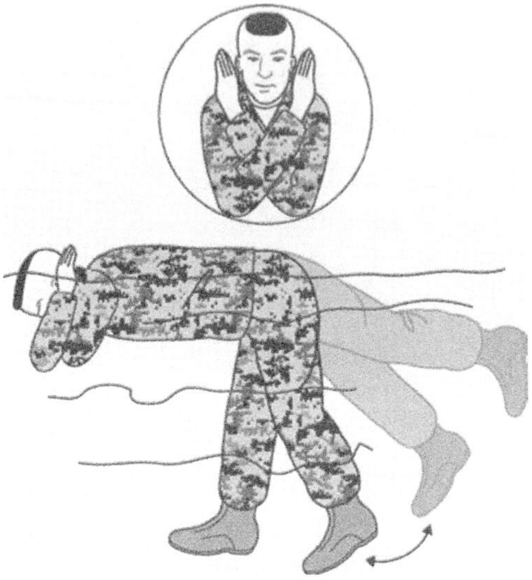

Bring your legs together and sweep your arms down and out until the arms are fully extended out to the sides. This raises your face above the water and allows you to catch a breath of air.

With your air supply replenished, return your face to the water and relax while sweeping your arms in a downward motion in front of your body to prevent/slow your descent. Do not hold your breath for more than 10 seconds. More than 10 seconds enhances your chances of shallow water black out and subsequent drowning.

Breast Stroke

Use this stroke to swim underwater, through oil or debris, and in rough seas. If you are a good swimmer and not wearing combat gear, the breast stroke is the best stroke for long-range swimming because it provides good visibility and allows you to conserve your energy and maintain a reasonable speed.

Body position

Lie prone in the water. Swim with your trunk and legs projecting back and down at an angle of 20 to 30 degrees.

Extend arms out in front (hands together [side by side]), and extend legs behind (toes pointed) to prevent drag.

Face downward, looking forward at a 45 degree angle to break the water and to prevent water from washing into the collar area causing drag.

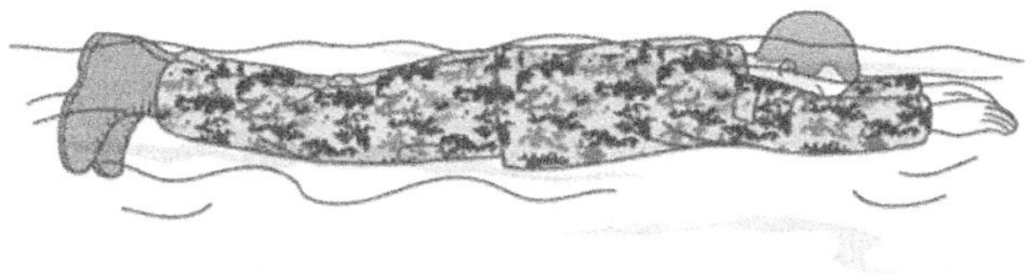

This is known as a glide.

Arm action

Turn your palms outward and bend your arms slightly.

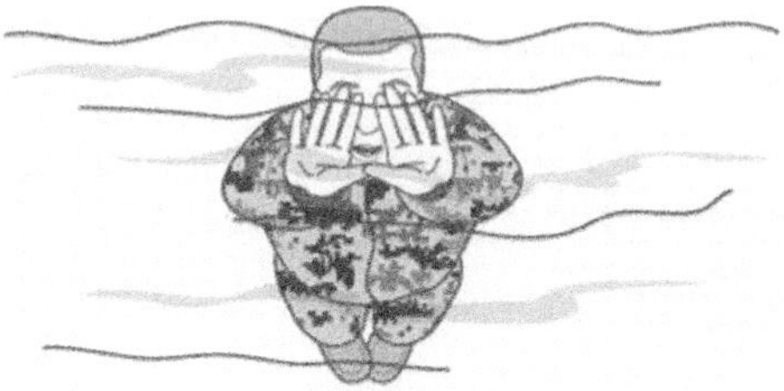

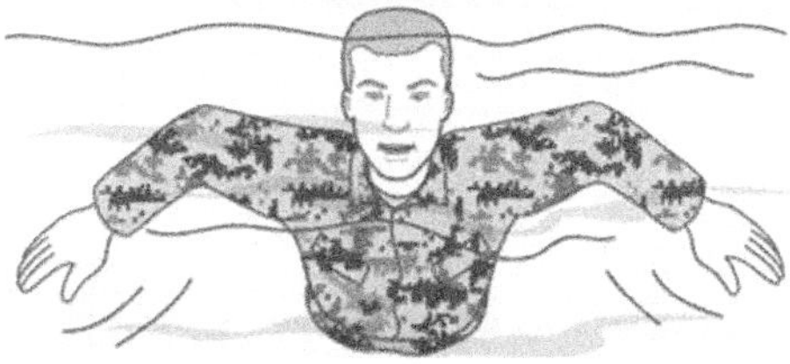

Sweep your arms sideward and slightly downward until your hands are opposite and slightly below your shoulders.

Rotate your head up, breathe once your mouth breaks the surface.

Bring your hands and arms up along your chest and thrust them forward until they are extended and ready to execute the next arm pull.

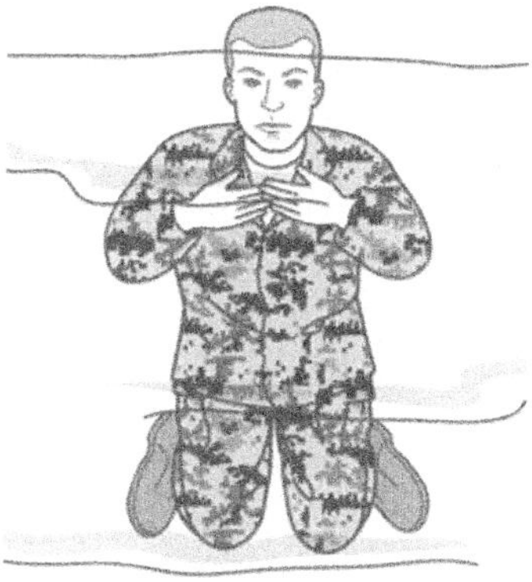

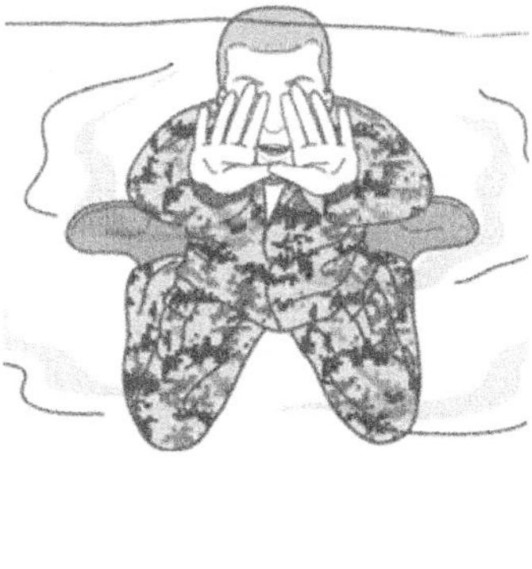

As the arms start their recovery into the glide, the head should rotate forward, resubmerging the face.

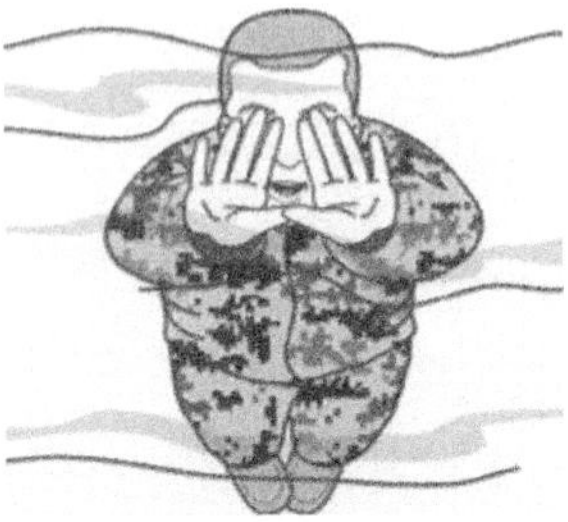

Leg action

Draw your heels toward your buttocks, establish a 45 degree bend in the knees.

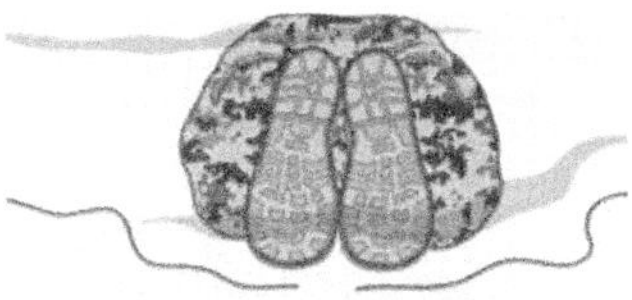

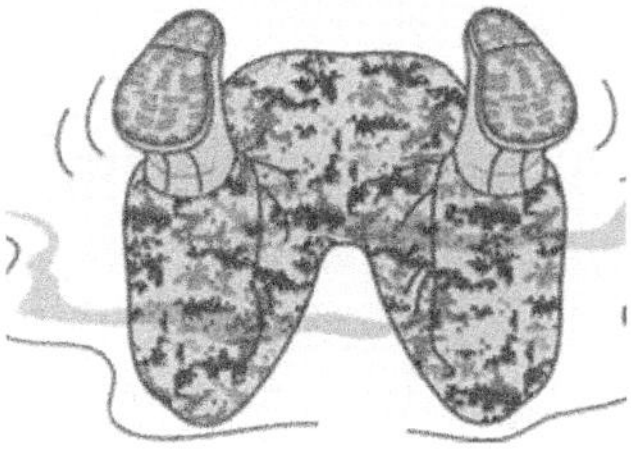

Thrust your legs outward and rearward, then squeeze them together. The whipping action of the feet aids forward propulsion.

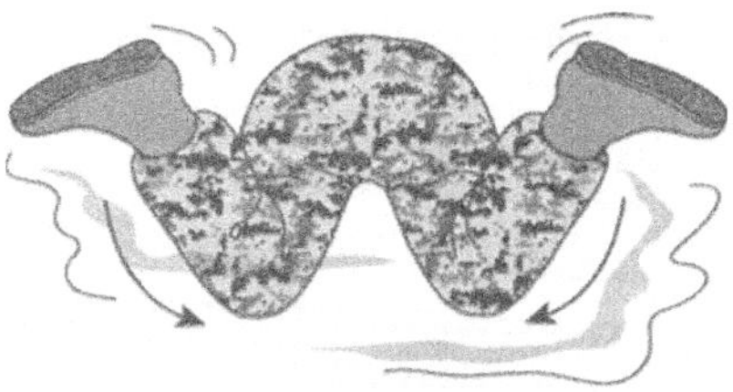

This is known as the breast stroke kick.

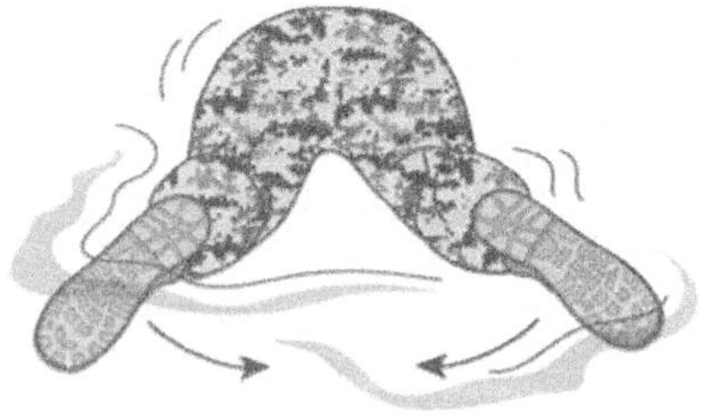

Breathing
Inhale during the arm pull and exhale through your mouth and nose during the finish of the breast stroke kick and glide.

Coordination
The stroke movement is in three counts:

Begin your arm pull. Near the finish of the pull, flex your knees and bring your heels toward your buttocks. The arm pull counteracts the resistance created by the knees.

As the arm pull is completed, thrust your hands forward, kick your legs outward and rearward, and squeeze them together.

Glide through the water for approximately 1 to 3 seconds or until your forward momentum decreases, then begin the next stroke.

Side Stroke

The side stroke is a survival stroke because you use both arms for buoyancy, with each arm creating a slight propulsion. The majority of your body's propulsion comes from your kick. To execute the side stroke—

Body position
Lie on your side with your lead (bottom) arm extended beyond (with a slight bend in your elbow) your head and in line with your body. Palm is down and your hand is submerged 6 to 8 inches.

Extend your trail (top) arm down the length of your body over your thigh.

Keep your legs straight and together, toes pointed rearward.

Keep your face out of the water, this allows for free breathing.

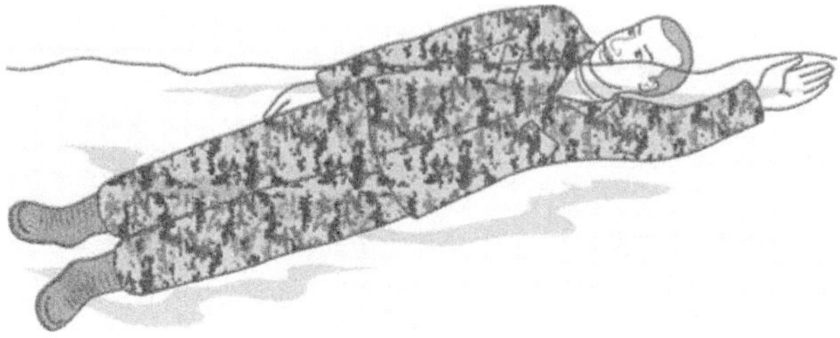

This is known as the glide.

Arm action

With your lead arm, pull your arm downward, while flexing at the elbow, until it is straight down from your shoulder.

Rotate your shoulder and pull your elbow into your side. This should put your lead hand at shoulder level. At the same time, turn your palm toward your face and thrust forward to your original, extended position.

Draw your right hand upward in front of your chest to shoulder level. Rotate your palm toward your feet, then push it downward in front of your body toward your feet to catch the water.

Push your trail hand backward to its original position on top of your thigh. (Your trail hand starts forward and meets your lead hand at your chest/shoulder.)

Leg action

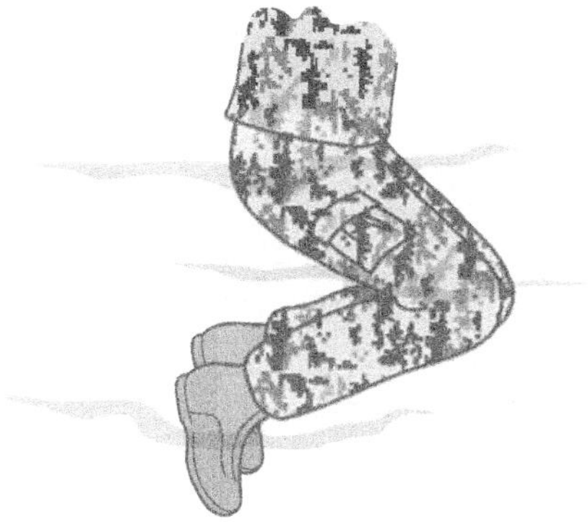

To perform the scissor kick, the top leg always goes forward and the bottom leg always goes rearward. From the extended position, draw—or recover—your feet toward your buttocks until your legs are bent at a 45 degree angle at the knees and the hips are flexed at a 45 degree angle with the thighs.

Once the legs have completed their recovery and while maintaining a 45 degree bend in the knees, extend the legs fully into a "V" shape in order to catch the water for the propulsion phase.

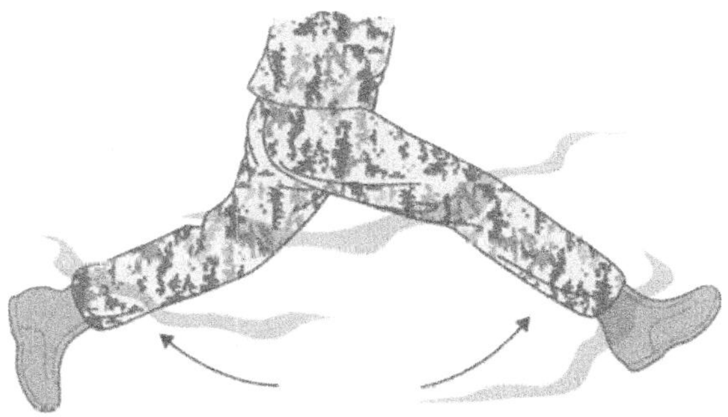

Once the legs are separated and extended forward and rearward to the "V" position, sweep the legs together until the feet are together.

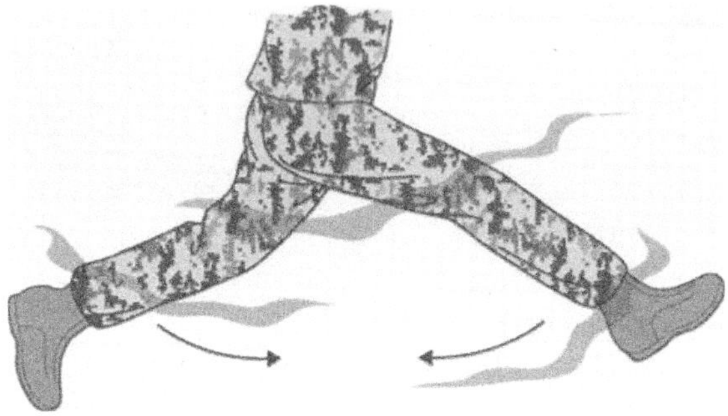

You are now in the glide position.

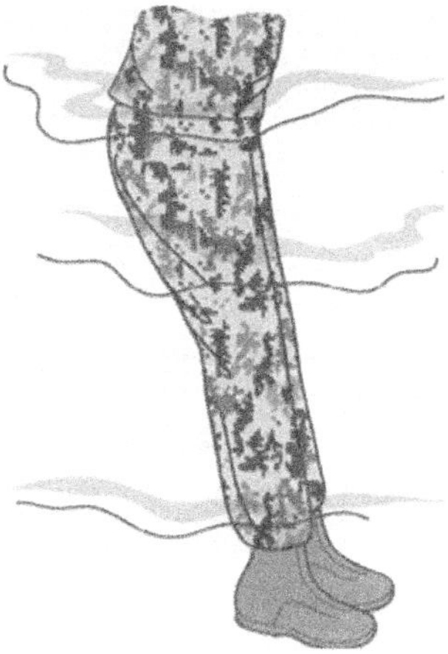

Breathing

As long as your face remains clear of the water, it is a free breathing stroke. However, it is recommended that you exhale then quickly inhale when the legs are sweeping back together in the scissor kick. This is when the body reaches its highest point in the water, thus clearing the face completely from the surface of the water making it the optimum time to breathe.

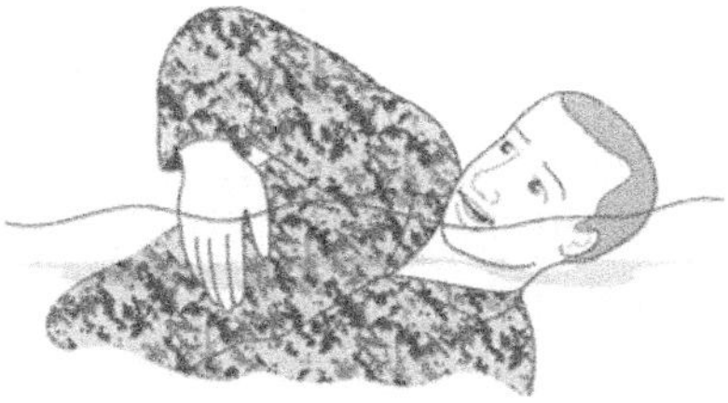

Coordination

Begin the stroke with the downward pull of your lead arm. At the same time, bring your trail arm upward and draw your knees up to begin the kick. Let the thrust of the lead arm, push of your trail arm, and the kick of your legs coincide in order to finish the glide position. Glide through the water for approximately 1 to 3 seconds or until your forward momentum decreases, then begin the next stroke.

Elementary Backstroke

The elementary backstroke is also an excellent survival stroke. It relieves the muscles that you use for other strokes, and it is the recommended stroke for weak swimmers or nonswimmers. To execute the elementary backstroke—

Body position
Start on your back.

Face up, chest up, and hips up, keeping an arch in your lower back with arms pressed to your sides and your legs extended and joined to prevent drag.

Arm action
Trace your hands up your sides to an area near your armpits then extend your arms out to the sides to form the letter "T" (palms facing feet), locking out the elbows.

> *Note:* Don't raise your arms above your head. This creates drag, changes your body position, and submerges the head.

Slap your palms to your thighs using a strong sweeping motion.

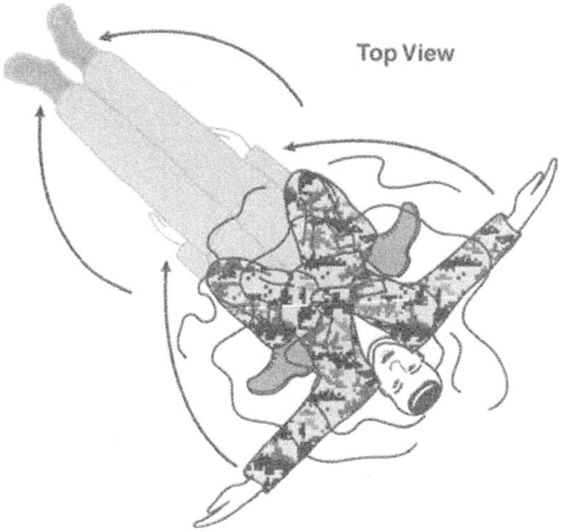

Leg action
Bend both legs at the knee (90 degree angle) slightly separating your knees and drawing your heels downward to a point under and outside your knees. The knees are spread as wide as the hips or slightly wider depending on the body type of the swimmer.

Circle around in a whipping action, ending with legs in a glide position.

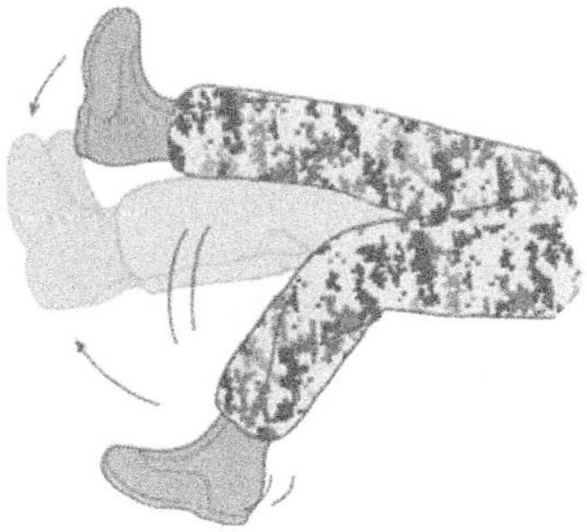

Breathing
Breathe anytime during this stroke. However, it is recommended that you exhale then quickly inhale when your arms are sweeping back toward your sides and while your legs are sweeping back together. This is when the body reaches its highest point in the water, thus clearing the face completely from the surface of the water and making it the optimum time to breathe.

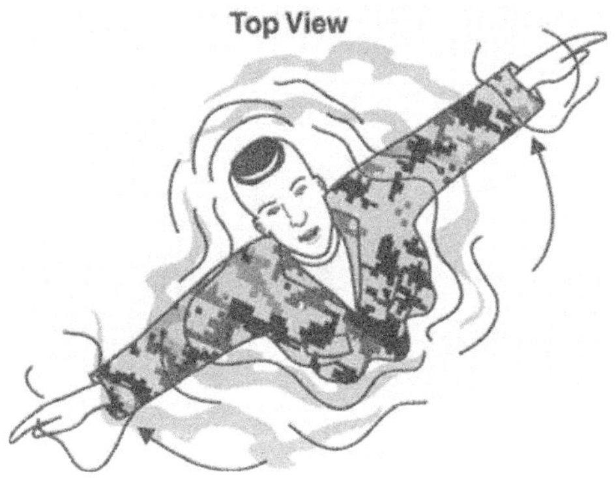

Coordination
The stroke movement occurs in three counts (recovery, catch, power).

Begin the arm pull (recovery).

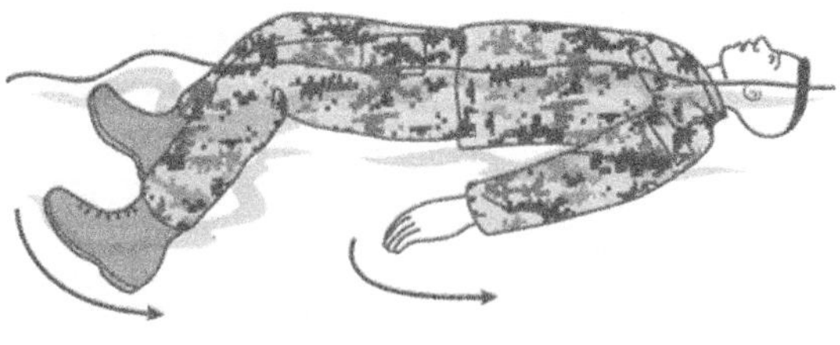

Near the finish of the pull, flex your knees to a 90 degree angle. The arm pull counteracts the resistance created by the knees.

Kick out your legs, and squeeze them together as the arm pull is completed (catch, power).

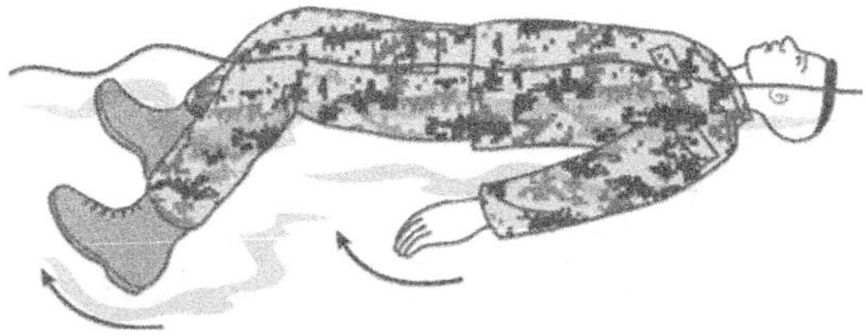

Glide through the water for 1 to 3 seconds or until your forward momentum decreases, then begin the next stroke as your momentum slows.

Chapter 2
Water Rescues

A drowning victim can panic and react with unexpected violence and can seize and inadvertently drown a rescuer. Therefore, if possible, a water rescue should be executed from a distance. Reaching, wading, or throwing methods are used in a distance rescue.

If a victim is too far away to use these methods, a swimming rescue may be necessary. Remove all combat gear, when possible, before entering the water. Swim within 2 to 6 yards of the victim to maintain a margin of safety, this allows you to reassess the situation and reassure the victim. If the victim is unconscious, use the wrist tow method or cross-chest carry method to pull the person to safety. If the victim is struggling, use a rear approach and then execute either a single armpit level off or a double armpit level off before towing the victim to safety.

If the victim does begin to overpower you, there are techniques that allow you to defend yourself without having to abandon the rescue. These techniques include the block, the wrist-grip escape, the front head-hold escape, and the rear head-hold escape. These techniques allow you to separate yourself from the victim, reassess the situation, and then attempt the rescue again.

> *Note:* In this chapter's illustrations, the rescuer is shown without a helmet.

Reaching Rescue Techniques

Reach

From a safe position at the water's edge, reach out to the victim. Talk constantly to calm the victim. Retain partial contact with land or some solid support structure (e.g., pier, bridge). If the victim is close but still beyond reach, extend an object (e.g., stick; pack; rifle with magazine removed, chamber empty, and muzzle pointing toward victim) that the victim can grasp. Pull the victim slowly to safety. Once the victim is close to shore, remove the victim from the water.

Note: You can also extend a foot to the victim if you can retain a secure grip on a solid support structure.

Reach From a Deck

Reach from a deck rescue can be executed by either a swimmer or a nonswimmer, and it can be used on an active or a passive victim. Be sure to reassure the victim during the rescue. To execute reach from a deck—

Lie prone on the deck with your body firmly anchored. To anchor your body, lie flat, spread legs apart, and extend one arm behind you with your palm down and on the deck.

Keep as much of your weight on the deck as possible and extend your free hand to the victim.

Grasp the victim's wrist from above, your thumb and index finger are facing you. The victim should never be allowed to grab you and put your life at risk. Therefore, never reach across the victim to grab his wrist, always grab the wrist that is the closest to you.

Keep your arm straight and locked out and pull the victim to the side of the deck. The victim should never be allowed to grab you and put your life at risk. Therefore, never pull the victim into you, always pull the victim into the side.

Arm Extension

An arm extension rescue can be used if you cannot reach the victim using the reach from a deck rescue technique and you must enter the water. An arm extension rescue technique is used for a victim who is either active or passive. Once you determine that a reach from a deck rescue technique is not viable—

Reassure the victim and quickly ease into the water while holding onto the deck with one hand.

Grasp the victim's wrist from above, your thumb and index finger are facing you. The victim should never be allowed to grab you and put your life at risk. Therefore, never reach across the victim to grab his wrist, always grab the wrist that is the closest to you.

Keep your arm straight and locked out and pull the victim to your side. The victim should never be allowed to grab you and put your life at risk. Therefore, never pull the victim into you, always pull the victim into the side.

Leg Extension

If the victim is beyond the reach of your arm, ensure that you have a firm grip on the deck, extend a leg to the victim and allow him to grab it, slowly bring the victim in closer until you can grab the victim's wrist that is holding onto your leg. Once you have a firm grasp on the victim's wrist, use the steps in the arm extension to pull the victim to safety. The leg extension rescue can only be used to rescue an active victim because the victim must be able to grab the rescuer's leg.

Wading Assist

Do not wade into water that is deeper than your chest. Talk constantly to calm the victim. If possible, do not touch the victim directly. Extend an object (e.g., stick; pack; rifle with magazine removed, chamber empty, and muzzle pointing toward victim) that the victim can grasp. Once the victim grasps the object, pull the victim slowly to safety.

Throw

If the victim is not within reach, use an expedient line to throw a lifesaving device to the victim. A lifesaving device can be any weighted item that floats (e.g., a canteen that is one fourth full of water). The lifesaving device must be secured to the end of the rope so that the rope will feed out from its coil when tossed. Talk constantly to calm the victim. Once the victim grasps the line or the lifesaving device, pull the victim toward you at a steady pace that keeps the victim's head above the water's surface. DO NOT pull so strongly that you break the victim's grip on the line. To prepare and use an expedient line and lifesaving device—

Tie a bowline at one end of the rope. (The appendix illustrates various types of knots.)

Unfasten the lid of a canteen.

Place the bowline around the neck of the canteen.

Refasten the lid so the canteen hangs from the bowline loop.

Place one end of the rope under the ball of your forward foot to secure it, either tie a knot at the end of the rope or tie an object at the end of the rope to create a block. Stand with your weight on the end of the rope. DO NOT tie the rope around your ankle.

Coil 20 to 30 yards of the rope, and hold it in a nonthrowing hand.

Place the canteen in your throwing hand.

Use an underhand throw to pitch the canteen and rope a short distance over the victim's head. Keep your nonthrowing hand open so the coil can unfold freely. The rope should trail across the victim's outstretched arms.

Retrieve the rope if the throw is inaccurate or the victim fails to grasp it. Recoil the rope as it is retrieved.

Divide the coil and throw again.

Lifesaving Approaches

Properly approaching the victim is as important as any other aspect of a rescue. Knowing the appropriate and safest method needed to approach a victim is critical in ensuring your safety and the victim's survival. Determining the victim's physical state (distressed swimmer, victim that is active and drowning, or victim that is passive and drowning) is crucial and will determine which type of approach you will execute.

Front Surface Approach

The front surface approach is typically performed when the victim is passive. You must remember that approaching the victim from the front is dangerous because a distressed swimmer or victim that is active can lunge toward you.

Rear Approach

If the victim is active, you must remember that his response can change rapidly. Therefore, approaching him from the front could be extremely dangerous. The victim could easily grab you and take you under the water during a state of panic. Therefore, use a rear approach when possible. When approaching a victim from the rear, reassure the victim until you get to within 2 to 6 yards of the victim; at that point, stop talking to the victim in order to maintain an element of surprise.

Approach Strokes

Before approaching a victim, you must evaluate the victim:

- Is the victim a distressed swimmer?
- Is the victim active and drowning?
- Is the victim passive and drowning?

Select your approach stroke based on your evaluation. Remember, the safety of the rescuer is paramount. Do not endanger yourself in an attempt to reach a drowning victim.

Crawl Stroke Approach Stroke

The crawl stroke (free style) is the fastest approach stroke. This stroke is used when the victim is passive and/or unconscious in the water. That is, the victim is face down, submerged or near the surface of the water, is not breathing, or is not moving. If any of these conditions exist, there is an obvious need to reach the victim as quickly as possible.

Execution of the crawl stroke approach stroke is the same as the crawl stroke (see pages 1-40 through 1-44), except that your head remains above the water's surface to allow for free breathing and maintaining eye contact with the victim.

Maintaining eye contact is critical. If the victim becomes submerged, you will have a better chance of locating him if you know his last location. Your distance from the victim will determine if your head remains out of the water the entire time you are executing the stroke. If you are 54 yards or less from the victim, you should maintain eye contact by keeping your head raised until you reach the victim. If you must swim more than 54 yards to reach the victim, make eye contact with the victim, place your face back in the water for four or five strokes, then raise your head again to regain eye contact with the victim (repeat these steps until you reach the victim).

Body position

The position of your body is horizontal to diagonal and influenced by the position of the head. This allows you to maintain eye contact with the victim.

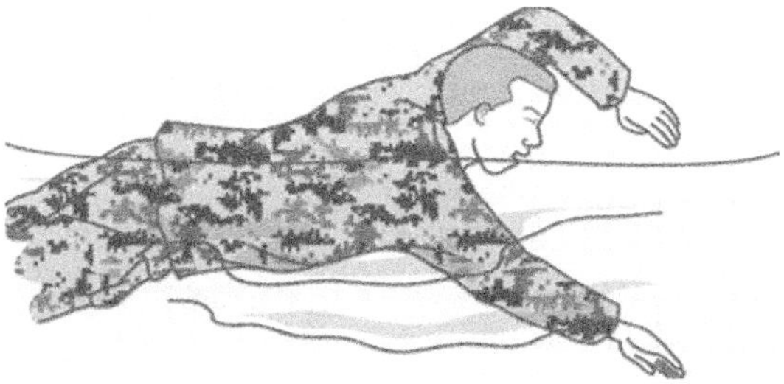

Arm action

While one arm is forward of your head to catch the water for the propulsion phase, extend your other arm alongside your body in what is known as the recovery phase. The recovery phase ends when your arm is extended forward of your head to begin the propulsion phase. In the propulsion phase, your arms are approximately shoulder-width apart and the pulling action is slightly wider and deeper to compensate for your raised head.

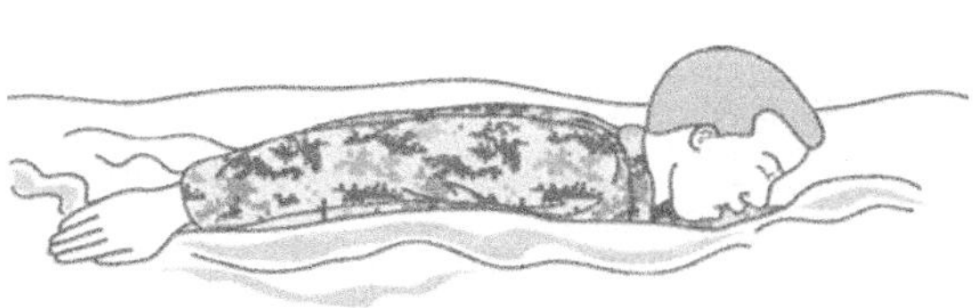

Leg action
Kick one leg in a downward motion while your other leg is recovering to the surface to prepare for the next kick (this is known as a flutter kick). This kick is used for propulsion and maintaining the lower body as horizontal to the water's surface as possible. When your head is raised, your knees are slightly bent in order to keep your feet near the surface.

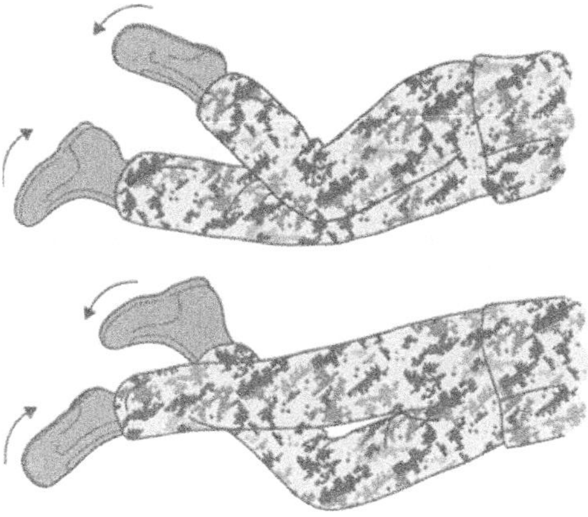

Breathing
If your face is above the water, breathe as needed (free breathing).

If traveling in excess of 54 yards to reach the victim, rotate your head to the right or left during the recovery phase and inhale when your face is out of the water or wait until your head is out of the water and facing forward while regaining sight of the victim.

Coordination
This stroke uses constant arm and leg action.

Breast Stroke Approach Stroke

The breast stroke approach stroke is used when a swimmer is distressed or a victim is active and drowning. This stroke is also used when the victim has a suspected spinal injury because it minimizes the wake created around the victim, minimizes the movement of the victim's head and neck, and helps prevent further injury.

If a victim is at a distance greater than 54 yards and if the swimmer is in distress or the victim is active and drowning in open water (ocean, lake), you should use the crawl stroke approach stroke to rapidly approach the victim, stopping approximately 11 yards from the victim in order to assess the situation.

Execution of the breast stroke approach stroke is the same as the breast stroke with a few minor deviations.

Body position
The position of your body is horizontal to diagonal and influenced by the position of the head. This allows you to maintain eye contact with the victim. It also allows you to communicate with the victim in order reassure him as you approach.

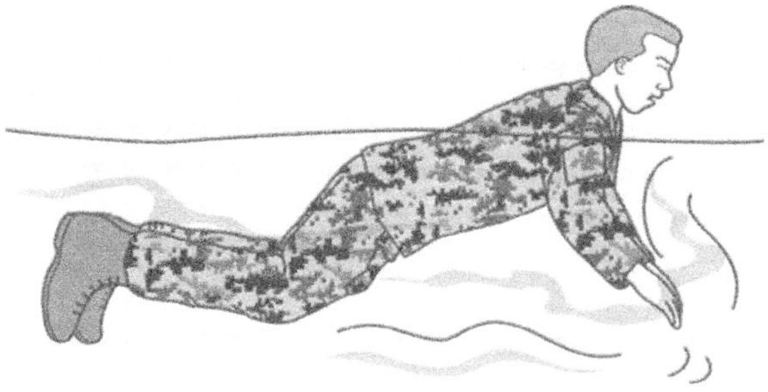

Arm action

Execute the arm action for the breast stroke, except that your arm pull is wider and deeper, which allows your head to remain above the water's surface.

Leg action
Execute the leg action for the breast stroke, except that your legs open wider during the propulsion phase of the stroke.

Breathing
Keep your face above the water and breathe as needed (free breathing).

Coordination
Coordination is the same as the breast stroke, except that there is no glide period. You constantly stroke until you reach the victim.

Level Offs

A properly performed level off positions the victim's face up and horizontal to the water's surface. There are three types of level offs: front surface approach, single armpit level off, and the double armpit level off. The single armpit level off and the double armpit level off are done by performing a rear approach, which is the preferred approach when a victim is active.

Front Surface Approach

The front surface approach is performed when the victim is passive. Approaching a victim from the front may place you in danger because a distressed swimmer or a victim who is active may lunge toward you. To execute a front surface approach—

Stop 2 to 4 yards from the victim to reassess the situation and reassure the victim.

Determine which arm to use in order to rotate the victim into a face-up position (your right hand on the victim's right wrist or your left hand on the victim's left wrist).

Move into position and prepare to grasp the victim.

Turn sideways and move toward the front of victim.

Once in position, reach forward and grab the victim's wrist. Your thumb is on the inside of the victim's wrist, as if you were checking his pulse with your thumb. Your remaining fingers wrap around the victim's wrist.

Lean back immediately and execute a powerful scissor kick or inverted scissor kick and perform short, vigorous pulls with your free arm.

As the victim begins to move forward in the water, kick, pull, and twist outboard on the victim's wrist. The momentum created from the kick and the pulling and twisting action of the victim's wrist will rotate the victim into a face-up and horizontal position in the water.

Extend and lock out your towing arm down the length of your body and execute a wrist tow to move the victim to safety.

Single Armpit Level Off

Approach the victim.

Stop 2 to 6 yards from the victim to reassess the situation, reassure the victim, and maintain your margin of safety.

Approach the victim slowly and grab his armpit with your hand (your right hand to victim's right armpit or your left hand to victim's left armpit). Position yourself sideways to the victim and place your elbow in the center of his back.

Pull with your hand that is in the victim's armpit while pushing with your elbow that is in the victim's back to place the victim horizontal on the water's surface (face up). To assist in placing the victim horizontally, use your free arm to execute short, vigorous pulls and use your legs to execute a scissor kick or inverted scissor kick.

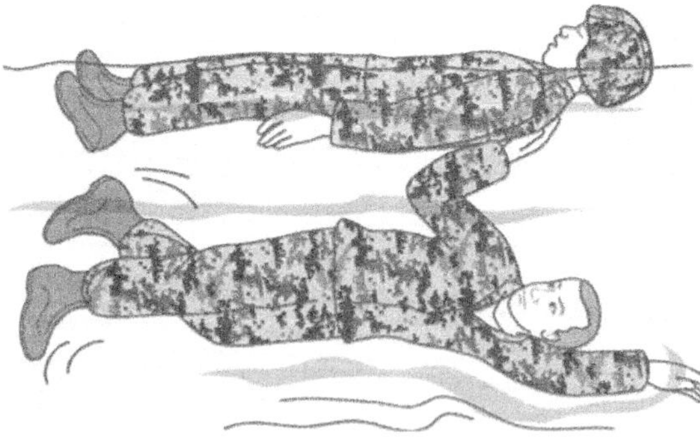

Once the victim is horizontal and on the water's surface, begin forward momentum by extending your towing arm to a fully locked-out position and executing a single armpit tow.

Double Armpit Level Off

Approach the victim.

Stop 2 to 6 yards from the victim to reassess the situation, reassure the victim, and maintain your margin of safety.

Approach the victim slowly and grab his armpits with both your hands. Place your elbows on his back.

Pull with your hands that are in the victim's armpits while pushing with your elbows that are in the victim's back and use an inverted breast stroke kick to place the victim horizontal on the water's surface (face up).

Once the victim is horizontal and on the water's surface, begin rearward momentum by slowly extending your arms into a fully locked-out position and executing a double armpit tow and inverted breast stroke kick.

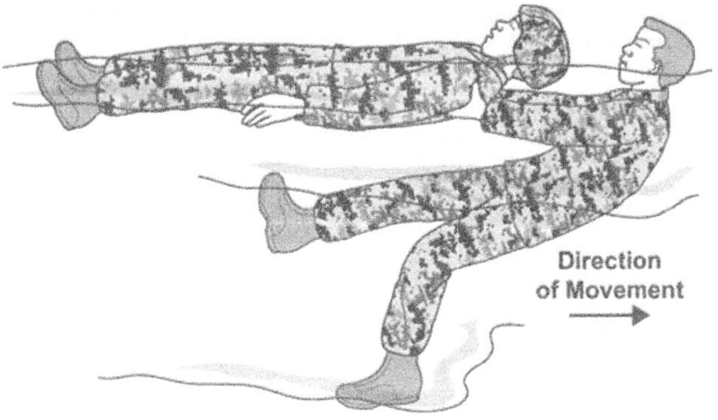

Rescue Techniques

Wrist Tow

Use the wrist tow method to rescue a victim who is floating face down. DO NOT use the wrist tow on a struggling victim. If time allows, remove your helmet and gear before attempting the rescue. Swim toward the victim using a modified breast stroke. Swim within 2 to 6 yards of the victim to maintain a margin of safety, this allows you to reassess the situation and reassure the victim. The following steps show proper front surface approach, wrist tow procedures:

Approach the victim from the front and grasp the underside of the victim's left wrist with your left hand or the right wrist with your right hand. Ensure that your thumb is on the underside of victim's wrist.

Lean back, pulling and kicking strongly to move the victim into a horizontal position.

Twist the victim's wrist to rotate the victim into a face-up position.

Swim toward safety using the lifesaving stroke.

> NOTE: The lifesaving stroke is a modified side stroke wherein the top arm is used to tow or carry a victim to safety and use a scissor kick or an inverted scissor kick is used for propulsion.

Keep a firm grip on the victim's wrist.

Keep your towing arm fully extended and along your side. This keeps the victim in the water column and prevents drag.

Ensure the victim's head does not go under water during the recovery.

Single Armpit Tow

You perform the single armpit tow after a victim has been properly leveled off. The single armpit tow uses the lifesaving stroke. To execute the single armpit tow—

Place one hand (top arm) in the victim's armpit. Your towing arm is straight and along your side.

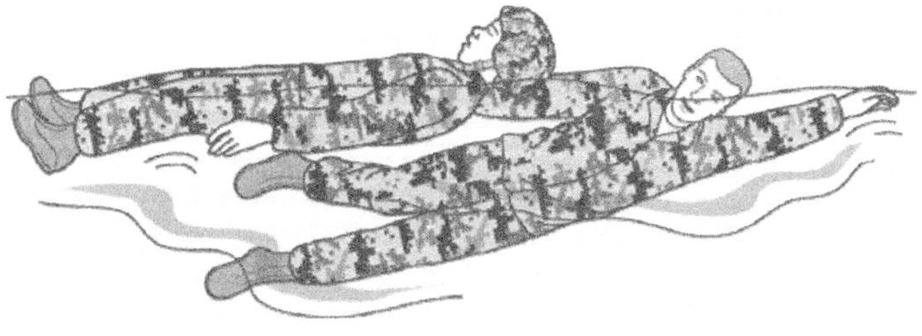

Use your lead arm (bottom arm) to execute short, vigorous pulls.

Execute a scissor or inverted scissor kick in a continuous and vigorous manner.

Use either free breathing or explosive breathing during the tow. During free breathing, keep your head above the water's surface and continuously reassure the victim. During

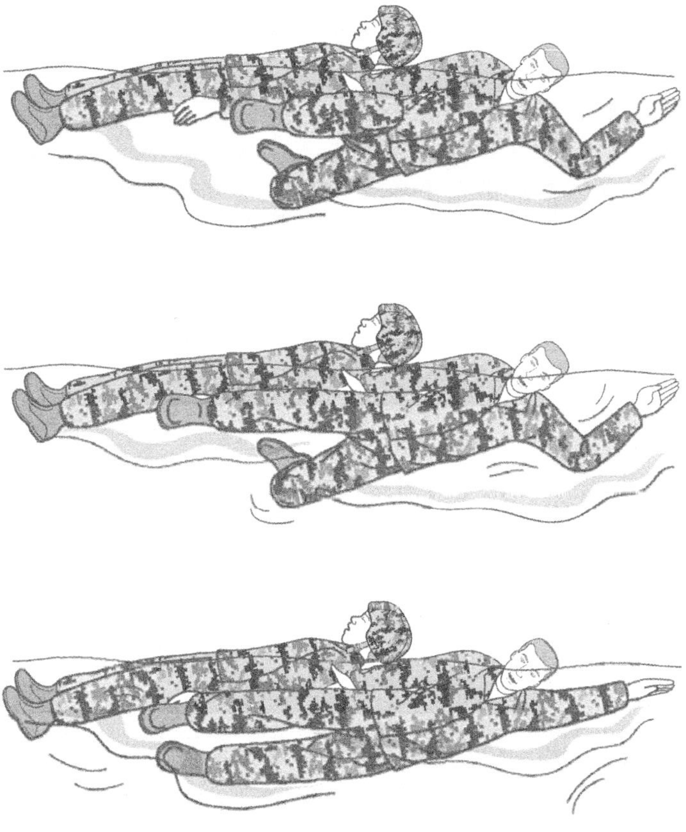

explosive breathing, put your head under the water to plane off your body's angle and reduce drag. Execute approximately two strokes, then lift your head up to breath and reassure the victim. Return your head back underwater for the next few strokes.

Double Armpit Tow

You perform the double armpit tow after a victim has been properly leveled off. To execute the double armpit tow—

Place both hands in the victim's armpits.

Extend both arms fully along your body. You are on your back, and your face is clear of the water.

Use free breathing.

Reassure the victim at all times.

Use the inverted breast stroke kick (it is the only kick that can be used due to your body position) to tow the victim to safety. (The inverted breast stroke kick is the same kick used in the elementary backstroke.) Your kick must be continuous and vigorous in order to keep the victim's face above the water.

Collar Tow

You perform the collar tow after a victim has been properly leveled off. To perform a collar tow using the victim's blouse—

Maintain control of the victim by grasping his armpit with one hand and then with your free hand grasp either his combat gear or his blouse between his shoulder blades. If grasping the blouse, grasp the material with your palm up, then turn your hand over to tighten the material.

Release the victim's armpit once control is established and execute the lifesaving stroke to tow the him to safety.

Cross-Chest Carry

Use the cross-chest carry to carry a victim to safety if the victim is struggling or when moving through heavy surf. Remove your helmet and gear before attempting a rescue. Talk constantly to calm the victim. Swim toward the victim using a breast stroke approach stroke. Swim within 2 to 6 yards of the victim to maintain a margin of safety, this allows you to reassess the situation and reassure the victim. The following steps illustrate proper cross-chest carry procedures:

> **CAUTION:** The cross-chest carry causes fatigue even if you are in excellent physical condition.

Use a level-off technique to place the victim in a horizontal, face-up position.

Retain a grip on the victim with one hand. Reach over the victim with your free hand to encircle the victim's chest. Place your free hand on the victim's opposite rib cage, just below his armpit.

Release your grip once you have a secure hold on the victim's chest.

Swim toward safety using the lifesaving stroke while keeping a firm grip on the victim's chest and your hip on his back.

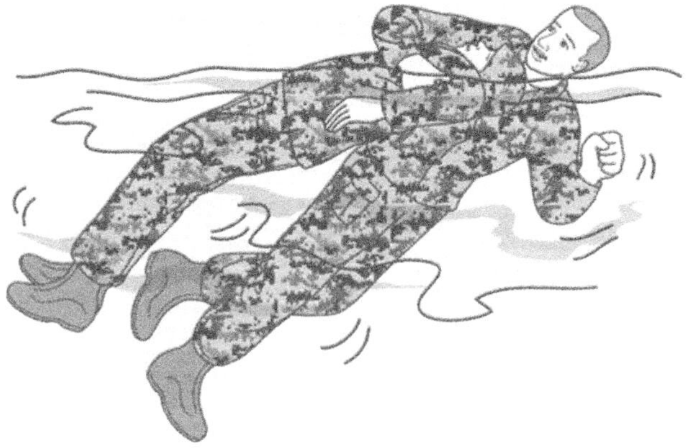

This procedure brings the victim's face and shoulders clear of the water, and typically the victim stops struggling. Sometimes, however, the victim will struggle during the swim to safety. If this happens, either tighten your grip on the victim or defend yourself with one of the techniques discussed on pages 2-25 through 2-30.

Tired Swimmer's Assist

You must maintain a 2 to 6 yards margin of safety from the victim at all times while you are getting into position to perform the tired swimmer's assist. Once behind the victim, you extend one arm, hand straight. Place your straight hand underneath the victim's opposite armpit. Maintain a 45 degree angle away from the victim, your arm is locked out. Assist the victim in propulsion until both of you reach safety.

Defense Against a Drowning Victim

DO NOT sacrifice your life in an attempt to save the victim. A struggling drowning victim poses great danger to anyone nearby. Driven by panic, the victim can grab you with great strength in an effort to climb out of the water. This can result in death for both you and the victim. The following techniques can be used to defend against a drowning victim, but the best defense against attack by the victim is to stay out of reach.

Block

The block prevents the victim from grabbing you if you have approached from the front. If the victim lunges toward you, react as follows:

Place one or both open hands against the victim's upper chest; being careful to avoid the victim's face, neck, and abdomen.

Lean backwards and submerge rapidly. Keep your blocking arm(s) extended.

Swim underwater and away from the victim, and return quickly to the surface.

Stop 2 to 6 yards from the victim to reassess the situation.

Determine an appropriate course of action.

Wrist-Grip Escape/Wrist-Grip Escape Alternative

The wrist-grip escape is used when a victim grabs your arm or wrist. Submerge the victim quickly by reaching across with your free hand, pushing down on the victim's shoulder to submerge him, and kicking to propel yourself upward. While keeping the victim submerged with your hand on his shoulder, give three hard jerking pulls with your trapped hand in an attempt to break free from his grasp. Once free, swim clear of the victim and reassess the victim's condition.

If the wrist-grip escape is unsuccessful in freeing your hand, use the wrist-grip escape alternative: use your free hand to grab your trapped fist, rotate thumbs up, apply bone-on-bone contact with the victim's arm, pry your hand out of his grasp, and quickly swim away from the victim.

Front Head-Hold Escape

The front head-hold escape allows you to escape when you are facing a victim who is gripping you around your head and neck. To execute the front-head hold escape—

Take a quick breath and tuck your chin into your shoulder to protect your throat.

Clap your hands above your head (three times) to submerge instantly. This drags the victim below the water, lifts his arms from around your neck, and, typically, he releases his grasp in order to get back to the surface. If he doesn't release his grasp, apply pressure to the victim's brachial pressure points (which are located inside of the upper arm, above the elbow).

Thrust the victim's arms up and away.

Keep your chin tucked to protect your throat, and swim underwater away from the victim and return quickly to the surface at the ready position.

Stop 2 to 6 yards from the victim to reassess the situation.

Determine an appropriate course of action.

Rear Head-Hold Escape

The rear head-hold escape allows you to escape when a victim is gripping your head and neck from the rear. To execute the rear head-hold escape—

Take a quick breath. and tuck your chin down, turn your head to either side, and raise your shoulders to protect your throat.

Take a strong stroke, clap your hands above your head (three times), and submerge instantly. This drags the victim below the water and, typically, he releases his grasp in order to get back to the surface. If he doesn't release his grasp, apply pressure to the victim's brachial pressure points (which are located inside of the upper arm, above the elbow).

Thrust the victim's arms up and away.

Twist your head and shoulders until free.

Swim underwater away from the victim and return quickly to the surface.

Stop 2 to 6 yards from the victim to assess the situation.

Determine an appropriate course of action.

Administering First Aid/Rescue Breathing

To administer first aid/rescue breathing, remove the unconscious victim from the water when possible and stop any bleeding.

First, open the airway and check for breathing. If the victim is not breathing, give two breaths of air. Check for a pulse. If there is a pulse but the victim is not breathing, continue rescue breathing. If there is no pulse, start cardiopulmonary resuscitation (CPR). For detailed information on administering CPR and rescue breathing, refer to American Red Cross CPR skill cards.

WARNING

> If the victim has no pulse and is not breathing, administer CPR immediately. If the victim does have a pulse but is not breathing, give rescue breathing only. If the victim has a pulse and is breathing, DO NOT give CPR—CPR could prove fatal.

Next, protect any wounds from exposure, and treat for shock.

Chapter 3
Treatment of Casualties and Avoidance of Dangerous Marine Life

To survive in the water, you will face many challenges other than just being able to swim. For example, injuries and/or fatigue can lead to drowning and exposure to the elements can lead to hypothermia or heat injuries and may require medical attention. Finally, your presence in the water may be seen as a threat to or food source for a variety of marine life and you may be attacked. This chapter advises you on how to minimize these risks in order to stay alive.

Drowning

Drowning is suffocation by a liquid. A drowning victim inhales water into the lungs or the throat closes by reflex so that little or no water can enter the windpipe. In either case, a victim can no longer breathe.

Symptoms

One symptom of drowning is that the victim may call for help and has an expression of dread or panic. But typically a victim that is active and drowning may not call for help because he is trying to conserve his air and will not speak.

Another symptom of drowning is when the victim thrashes at the water's surface. If the victim stops or grows calmer, he has likely been overcome by fatigue, hypothermia, or a lack of air. At this stage, the victim usually has 20 to 60 seconds before going under the water's surface.

Treatment

If the victim is not breathing, begin rescue breathing. Place the victim on his back, tilt head back to open airway, pinch the nose, and give two full breaths. If the victim does not inhale during the first two breaths, reposition his head and attempt two more breaths. Check for a pulse. If a pulse is present, but the victim is still not breathing, continue rescue breathing. If a pulse is not present, begin CPR. See MCRP 3-02G, *First Aid*, for rescue breathing and CPR details.

WARNING

If the victim has no pulse and is not breathing, administer CPR immediately. If the victim does have a pulse but is not breathing, give rescue breathing only. If the victim has a pulse and is breathing, DO NOT give CPR—CPR could prove fatal. Continue first aid until medical help arrives.

A victim who is not breathing and has no pulse may appear dead. DO NOT decide that death has occurred. Continue with the prescribed treatment. A corpsman or medical officer should decide whether the victim can or cannot be revived.

Hypothermia

Hypothermia is the abnormal lowering of the body's internal (or core) temperature to 95 degrees or below. It occurs when the body loses heat faster than it produces heat. A hypothermia victim loses the ability to move quickly, becomes mentally sluggish, slips into semiconsciousness, lapses into a coma, and dies when internal body temperatures drop too low.

The chilling effects of cold air, wind, or water can produce hypothermia. Water poses the greatest threat because it transfers heat 25 times faster than air. Depending on the water's temperature, a victim can succumb to hypothermia within a few minutes. The body's sudden contact with cold water can also set off a body reaction known as the mammalian diving reflex. This reflex can greatly increase survival time (especially for women and children) in or under cold water. The mammalian diving reflex shuts off blood circulation, except for the flow between the heart, lungs, and brain. The small amount of oxygen left in the blood and lungs is saved for the body's vital organs. This reflex has allowed people to survive being under cold water for an extended period of time. Therefore, a cold water drowning victim should be treated as if still alive even though the victim is not breathing, has no pulse, and may appear dead. DO NOT decide that death has occurred. Continue with the prescribed treatment. Victims of hypothermia can appear to be dead when they are not. A corpsman or medical officer should decide whether the victim can or cannot be revived.

Warning

If the victim has no pulse and is not breathing, administer CPR immediately. If the victim does have a pulse but is not breathing, give rescue breathing only. If the victim has a pulse and is breathing, DO NOT give CPR—CPR could prove fatal. Continue first aid until medical help arrives. Check for a pulse for at least 45 seconds.

Symptoms

Once the body's core temperature drops, the victim will show one or more of the following symptoms:

- Violent and uncontrollable shivering as the body tries to warm itself.
- Slow or slurred speech.
- Disorientation or poor coordination.
- Loss of skin color.
- Blue and pinched lips.
- A slowing or stopping of shivering that progresses into a rigid torso and limbs.

Survival Time

A hypothermia victim's survival depends on the water's temperature and the time spent in the water. A small body build cools faster than a large build. Children cool faster than adults.

To increase your chance of survival in the water, utilize the HELP position described on page 1-37. Extra clothing and inactivity (remaining motionless in the water) can also increase your survival time.

Treatment

A hypothermia victim must be warmed to prevent further heat loss; therefore, treatment should begin as soon as possible. Consciousness of the victim determines the treatment that should be pursued. See MCRP 3-02G for specific treatment information. The following treatment procedures are recommended:

- If the victim is conscious, give the victim warm fluids. Give candy or sweetened foods to a victim who is able to eat.
- If the victim is unconscious, place him on his back with his head tilted back to ensure an open airway.
- DO NOT massage the victim. Massage can break blood vessels and create swelling, internal pressure, and blocked blood circulation.
- DO NOT give alcohol to the victim. Alcohol lowers the victim's body temperature.
- Shock is a possibility, treat accordingly.
- Seek medical help immediately.

If you are able to remove the victim from the water, apply the following steps when possible:

- Get the victim into shelter.
- Remove the victim's wet clothing.
- Put the victim in dry clothing.
- Place the victim in a sleeping bag if one is available. It may be necessary to place another Marine in the sleeping bag with the victim.
- Place as much insulation as possible between the victim and the ground.
- Use hot water bottles, electric blankets, or blankets heated in an oven or by a campfire to warm the victim's neck, groin, and the sides of the chest.

CAUTION: DO NOT apply heat to extremities.

Heat-Related Injuries

Heat-related injuries include heat cramps, heat exhaustion, and/or heat stroke. They can occur when a Marine—

- Is exposed to extreme heat, such as from the sun or a combination of high air temperatures and water temperatures.
- Does not wear proper clothing or gets overexposed to the sun.
- Becomes dehydrated.

With heat cramps, muscles may cramp in the arms, legs, and/or stomach. The victim may sweat excessively. To treat, create improvised shade for the victim and have him drink water.

With heat exhaustion, the victim sweats heavily; presents pale, moist, cool skin; and complains of a headache, weakness, dizziness, and/or loss of appetite. A victim may also experience heat cramps, nausea, vomiting, an urge to defecate, chills, rapid breathing, confusion, and a tingling sensation in the hands or feet. To treat, pour water over the victim and fan him to speed up the coolant effect of evaporation, have the victim drink freshwater, and attempt to provide shade.

With heat stroke, the victim stops sweating, which results in red, flushed, hot, dry skin. The victim may first experience headaches, dizziness, nausea, fast pulse and respiration, and/or seizures and mental confusion. The victim can become unconscious and die if not treated quickly. To treat, create improvised shade for the victim. Have him drink water and elevate both legs if possible.

Burns

Apply a field dressing (or the cleanest material available) to the burn. Give sips of water to a casualty who is conscious and not nauseated. When treating burn casualties—

- DO NOT remove clothing stuck to the burns.
- DO NOT break any blisters.
- DO NOT apply grease or ointments to the burns.

For electrical burns, check for both an entry and exit burn from the passage of electricity through the body. An exit burn may appear on any area of the body, not necessarily opposite the entry burn.

For burns caused by wet or dry chemicals, flush the burns with large amounts of water and cover with a dry dressing.

For burns caused by white phosphorus, flush the area with water, then cover with a wet material, dressing, or mud to exclude air and keep the white phosphorus particles from burning.

For laser burns, apply a field dressing.

Common Medical Problems Associated with Sea Survival

Seasickness

Seasickness is the nausea and vomiting caused by the bobbing motion created by the wave action of a flotation device. Seasickness can result in—

- Dehydration and exhaustion.
- A loss of the will to survive.
- Others becoming seasick.
- Unclean conditions.

To treat seasickness—

- Wash both the Marine and the flotation device to remove the sight and odor of vomit.
- Keep the Marine from eating food until the nausea is gone.
- Have the Marine lie down and rest.
- Give the Marine seasickness pills if available. If the Marine is unable to take them orally, the pills should be inserted rectally for absorption by the body.

Saltwater Sores

Saltwater sores occur when skin that has abrasions or is cut is exposed to saltwater. The sores may form scabs and pus. Do not open or drain. Flush the sores with freshwater, if available, and allow to dry. Apply antiseptic, if available.

Blindness/Headache

Irritants or the effects of the sun's rays reflecting off the water can cause temporary blindness or headaches. If flames, smoke, or other irritants get in your eyes, flush the eyes immediately with saltwater, then with freshwater, if available. Apply an ointment, if available. Bandage both eyes for 18 to 24 hours or longer if the damage is severe. If glare from the sky and water causes your eyes to become bloodshot and inflamed, bandage the eyes lightly. Try to prevent this problem by wearing sunglasses or goggles with a sunglass insert.

Constipation

This condition is a common problem associated with dehydration. For constipation, do not take a laxative if it is available; this causes further dehydration. Drink freshwater, if available.

Sunburn and Dehydration

The sun's rays reflect at all angles off the waves of the water; therefore, sunburn and dehydration are serious problems in sea survival. Try to prevent sunburn by—

- Erecting an improvised canopy, with available floating materials, to provide shade.
- Wearing your soft cover or using a cloth, such as a handkerchief, to cover your head.
- Covering your skin with sunscreen or lip balm from your first aid kit. Your lips, nostrils, eyelids, the backs of your ears, and the skin under your chin sunburns easily. If enough sunscreen cream is available, all exposed skin should be covered.

Dehydration is caused by the loss of the body's vital fluids. Dehydration in saltwater may result from a combination of factors such as a lack of water, the effects of saltwater on skin tissue, sunburn or vomiting from seasickness, and other causes. Sleep and rest and reduced water and food intake are the best ways of enduring periods of exposure. The following measures will delay the effects of dehydration:

- DO NOT drink saltwater.
- DO NOT drink urine.
- DO NOT drink alcohol.
- DO NOT smoke.
- DO NOT EAT unless water is available.

Dangerous Marine Life

You may see many types of marine life around you, and some are more dangerous than others. Generally, sharks are your greatest danger, followed by barracudas. However, most marine life will not deliberately attack a human. The most common injuries from marine life are wounds from bites, stings, or punctures. With the exception of sharks and barracudas, most injuries are a result of either trying to catch game or from contact abrasion with marine life. To treat an injury that results from a dangerous marine animal bite—

- Pack the wound with gauze and then apply a pressure bandage, if available.
- Treat for shock.
- Obtain medical attention as soon as possible.

For a more detailed information on the treatment of casualties and dangerous marine life, see MCRP 3-02F and MCRP 3-02G.

Sharks and Barracudas

Only about 20 percent of all shark species are known to attack people. Sharks have an acute sense of smell, and the smell of blood in the water will draw them to their prey. They are also very sensitive to any abnormal vibrations in the water; therefore, the sound caused by a struggling swimmer or underwater explosions will attract them.

Barracudas are bold and inquisitive fish. They have been known to attack men who are wearing shiny objects. Barracudas may charge at lights or shiny objects at night.

A group of swimmers can maintain a 360 degree watch while in the water. Therefore, to protect yourself from sharks and barracudas, stay with other swimmers. The group members can either frighten away or fight off sharks or barracudas better than one person can. Keep all your clothing on, including footwear. Historically, sharks attack unclothed individuals in groups first, mainly in the feet. Clothing also protects you against abrasions from a shark's tough skin should the shark brush up against you. Avoid urinating heavily, let urine dissipate between discharges. If you must defecate, do so in small amounts and toss it as far away from the group as possible. Do the same if you must vomit.

If attacked, the use of firearms by swim sentries should be used with extreme caution because of the risk of injury to other swimmers. If unarmed or unable to make an improvised weapon, kick and strike the shark. Avoid using your bare hands to strike the shark, injury can result to your hands due to a shark's tough skin. Target areas on a shark are the gills, eyes, and underbelly. Blows to the snout are also not recommended because a shark will tilt its head up and thrust its jaws forward when biting.

Sea Snakes

Sea snakes are venomous and sometimes found in mid-ocean. They are unlikely to bite unless provoked. Your best protective measure is to avoid them.

Poisonous Fish

Many reef fish are toxic, and can kill you if eaten. Generally, poisons are present in all parts of the fish, but especially concentrated in the liver, intestines, and eggs.

Turtles and Moray Eels

Turtles and moray eels normally inflict minor bite wounds. Treat this type of injury by cleaning the wound.

Corals

Coral, dead or alive, can inflict painful cuts. Clean all coral cuts thoroughly. DO NOT use iodine to disinfect any coral cuts. Some coral polyps feed on iodine and may grow inside the flesh.

Jellyfish, Portuguese Man-of-War, Anemones, and Others

This group of marine animals inflicts injury by stinging their victims with their tentacles. Contact with their tentacles produces burning pain, a rash, and small hemorrhages on the skin. Shock, muscular cramping, nausea, vomiting, and respiratory distress may also occur. Gently remove the clinging tentacles with a towel and wash or treat the area. Use diluted ammonia or alcohol and talcum powder to treat the injury if available.

Spiny Fish, Urchins, Stingrays, and Cone Shells

These animals inject their venom by puncturing the skin with their spines. General signs and symptoms include swelling, nausea, vomiting, generalized cramps, diarrhea, muscular paralysis, and shock. Deaths are rare. Treatment consists of soaking wounds in hot water, if available, to deactivate heat-sensitive toxins.

Chapter 4
Negotiating Water Obstacles

Marines face water obstacles in saltwater, freshwater, and brackish water (where freshwater and saltwater meet). These water environments differ considerably and pose distinct problems for Marine tactical units and swimmers. Saltwater obstacles include tides, surfs, and currents. Brackish water obstacles include back bays. Freshwater obstacles include rivers and canals.

Tides

Tides are periodic changes in the surface levels of oceans, bays, gulfs, inlets, and rivers. The Moon's and Sun's gravitational pulls cause tides. Depending on the situation, tides can either help or hinder Marines in their endeavors to conduct amphibious or riverine operations. Tides can provide sufficient depths to allow for the passage of landing craft or boats over reefs, trees, and other underwater obstructions. On the other hand, tides can render a river fording site unusable to tactical vehicles. Direction, level of change, and amount of change determine tidal nomenclature.

Tides that show change in direction are flood tides and ebb tides. Rising tides are known as flood tides. Falling tides are known as ebb tides.

Tides that show extreme levels of change are high tides and low tides. High tide is the period when water is at its greatest depth. Low tide is the period when water is at its most shallow depth.

Tides that show the least and most amount of change are neap tides and spring tides, respectively. Neap tides have the least amount of change in water levels between high and low tide. Neap tides occur at the half moon, when the Sun and the Moon are aligned at a 90-degree angle with the Earth. In this position, the Sun's and Moon's gravitational pulls offset each other. Spring tides have the highest floods and lowest ebbs. Spring tides occur at or shortly after the new moon or full moon when the Sun, Moon, and Earth are approximately in line. In this position, the Sun's and Moon's gravitational pulls are combined.

Surf

Waves break upon entering shallow water and create surf. The offshore area where waves break is the surf zone, which can present many hazards. Breaking waves often trap air bubbles and create a foamy appearance. Bubbles lower the water's density and decrease buoyancy. Move through foamy surf as quickly as possible. The type of wave determines your survival technique.

Wave action moves you toward shore. Lie on your back or side with your head pointing in the direction of the beach and your feet pointing into the waves.

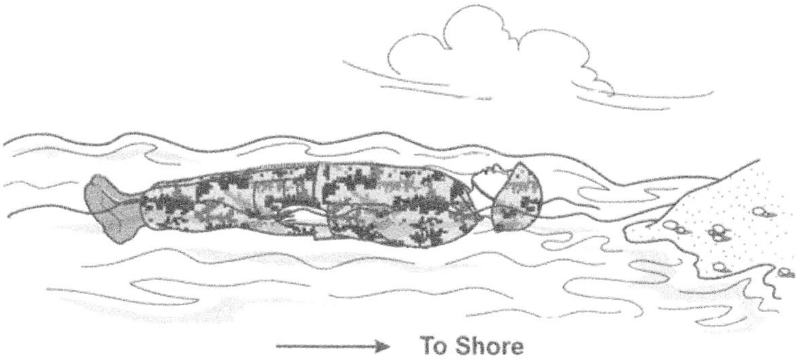

As one wave approaches the beach, another drains away; relax and do not swim against the draining water. When a new wave is within about 3 yards, start swimming toward shore. Continue to swim until the wave lifts you and moves you toward the beach. Once the wave loses forward momentum, relax and repeat the cycle. If nearing rocks, turn your body and approach feet first to reduce the chance of striking your head and arms.

Plunging Waves

A plunging wave is a breaker whose crest curls forward and falls ahead of its base. Because of its power and underwater turbulence, a plunging wave poses the greatest surf threat. If caught in a plunging wave, you can be pulled underwater and pitched about violently. This can cause you to panic, which can increase your chance of drowning. Perform the following steps to escape a plunging wave:

Tuck into a ball with your head against your knees and your forearms locked around your legs, just below the knees.

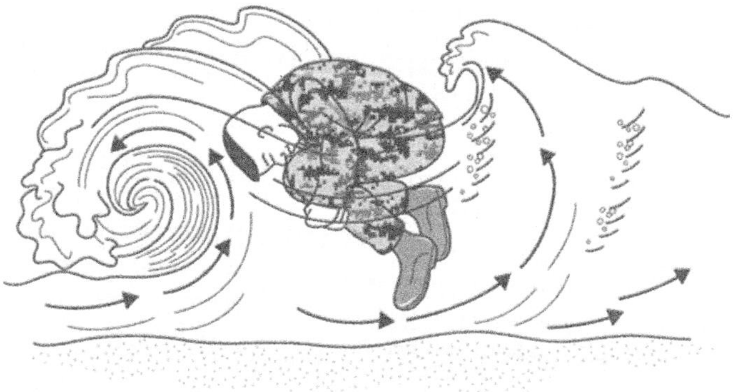

Relax in this position until the turbulence subsides and you float to the surface. This can take 30 seconds or more.

Swim toward shore.

Note: If threatened by another plunging wave, dive underwater into the wave.

Spilling Waves

A spilling wave does not break. Instead, its crest slides forward without curling. A spilling wave creates less turbulence and poses less of a threat than a plunging wave. If caught in a spilling wave, relax and let the wave carry you to shore.

Surging Waves

A surging wave occurs on a beach with a steep underwater gradient. It never really breaks, but the crest rises while the base slides up the beach with great force and speed. Once the wave reaches its highest point on the beach, it rushes back as quickly as it surged forward. If you are standing on the bottom when a surging wave advances or retreats, the wave can knock you off your feet and pull you into the surf zone. Do not try to stand or walk on the bottom. Swim toward the beach as soon as possible.

Currents

Offshore Currents

An offshore current occurs outside the surf zone. Typically, it occurs at bay entrances, in island channels, and between islands and the mainland. An offshore current flows parallel to or away from shore. If the offshore current is created by tides, its current strength and direction vary at different times of the day.

If caught in an offshore current, you may be carried in a direction you do not want to go. DO NOT try to swim directly to safety. If the current is moving directly away from the shore, relax and wait until the current dies out or turns toward land. Once the current subsides, use any survival stroke to swim toward shore. If the current is moving parallel to shore, use any stroke to move at an angle across the current and toward shore.

Rip Currents

A rip current occurs when waves pile water against the shore faster than the water can drain. The water flows rapidly along the beach until it is deflected seaward by a bottom obstruction. Then the rip current flows through the surf zone and into open water at a speed of up to 2 knots. This action can cut deep trenches in the sand. A rip current dies out once it reaches open water (usually within a few hundred yards of the shore).

A rip current can pull you out to the open sea. If caught in a rip current, DO NOT try to swim against the current. A rip current moves faster than most people swim, and it is impossible to swim to shore once caught in it. Relax and stay afloat until the current runs out. Once the current subsides, use a survival stroke to move parallel to the shore until you are out of the current, then begin swimming toward shore.

Littoral Currents

A littoral current occurs when a wave breaks against a beach at an angle. This current flows parallel to the shoreline and does not pose a great threat. If caught in a littoral current, use the combat travel stroke to swim across it at an oblique angle.

Back Bays

Once on the beach, you often face one or more rows of low hills called dunes. Behind the dunes, you may encounter a low-lying stretch of ground thickly covered with scrub trees and bushes. This area gives way to wetlands known as back bays. Back bays consist of muddy islands that are almost submerged during flood tides and separated by channels of brackish water of varying depths. Channel bottoms usually contain soft mud. Back bays pose major obstacles to vehicular traffic.

Infantry can cross back bays, but only with great effort. If crossing back bays by foot, consult detailed navigation charts and use the following guidelines to plan your route:

- Avoid water less than waist deep; walking in shallow water or soft mud is extremely tiring.
- Avoid back bay islands; these low-lying islands are usually too muddy to support foot traffic.
- Seek out deep water; floating with a pack is less tiring than walking through shallow water or soft mud.
- Seek out sand, shell, gravel, or stone bottoms; these firmer bottoms generally ease travel and help conserve energy.

Rivers and Canals

A river is a large, natural stream of water that empties into a larger body of water. The slope of the riverbed and the volume of water in the river determine its current.

Canals resemble small rivers or streams in their width and depth, but usually lack any significant current. Climbing out of these waterways can be difficult if the canal is flanked by steep banks.

Chapter 5
Fording Waterways

WARNING

Fords are dangerous. Cross as quickly as possible.

A ford is any site in a river, stream, or canal where the water is shallow enough for troops or vehicles to cross without using flotation devices. Canal bottoms are usually too soft to support fording vehicles, and wading infantry frequently stumble.

The tactical situation dictates the location of the fording site. Seek fords that are protected from enemy observation and that allow for adequate supporting fires. A night fording takes at least one and a half times as long as a daylight fording.

> *NOTE:* In this chapter, the term "river" refers to rivers, streams, and canals.

Selection of a Ford Site

The following table identifies desirable fording characteristics:

Characteristics	Comments
Concealment	The ford hides personnel and vehicle movement from enemy observation.
Accessibility	The ford should have low banks with gentle gradients. This allows a free flow of traffic at both the entrance and the exit.

Characteristics (Continued)	Comments (Continued)
Slow Current	The ford's current should not exceed 1.5 meters per second if possible.
Firm Footing	The ford's bottom, entry, and exit composition should be firm enough to support traffic. Do not drive a vehicle over any bottom composition that a 2-inch diameter stick can be pressed into more than 1 or 2 inches.
Gently Sloped Channels	The ford's entry and exit points should be gently sloped. If possible, locate a portion of the stream where the channel is not actively shifting.
Depth	The fording depth is less than or equal to the least capable vehicle.

Determine the Slope

Units move into and out of water faster and more quietly if entry and exit points are not steep or muddy. Slope is the amount of change in ground horizontal distance (run) and in vertical elevation (rise) from one point to another. Slope is usually expressed as a percentage. You can use a clinometer, map, or line of sight and pace to measure the percentage of a slope.

Clinometer

A clinometer measures percentage of a slope. It is a component of the M9 armored combat excavator, and organic to many Marine engineer units. However, engineers are also taught (and often use) field-expedient methods to determine slopes and gradients.

Map

A map measures the horizontal distance along a desired path. Determine the difference in elevation between the path's starting and ending points. From a map's scale, you can determine the distance between two points. From a map's elevation lines, you can determine the difference in height between the same two points. Both figures must be in the same unit of measure (e.g., feet, meter). Divide the elevation (rise) by the distance (run) and multiply by 100.

Rise = 165 - 120 = 45 meters

Run = 200 meters

$$\% \text{ Slope} = \frac{45}{200} \times 100 - 22.5\%$$

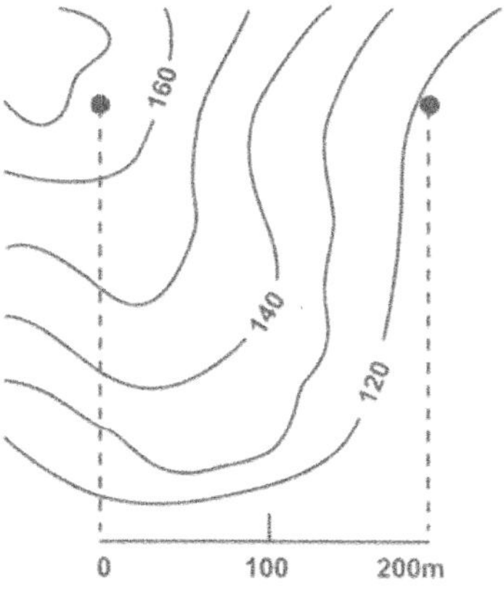

Line of Sight and Pace

To determine line of site and pace, stand at the bottom of the slope, keep your eyes level, pick a spot on the slope, then pace the distance. The number of paces multiplied by a standard measure of 0.75 meter determines the run. The eye-level height (usually 1.5 to 1.7 meters) determines the rise. Repeat this procedure until you have covered the entire distance you want to measure for each spot (vertical and horizontal). Add the vertical distances to provide total rise and the horizontal distances to provide total run.

Determine the Current Speed

Current speed increases as channels narrow. It may be necessary to locate a wider ford location to obtain a slower stream current. The following steps are used to calculate the speed of the current:

Determine points A and B along a channel. Then measure the distance between those two points.

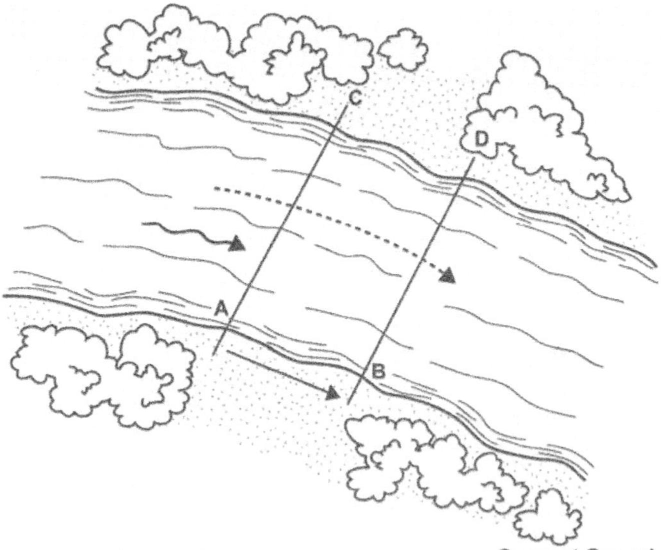

$$\frac{\text{Distance (meters) Between Points A and B}}{\text{Time (seconds) Between Points C and D}} = \text{Current Speed (meters per second)}$$

Sight directly across the water from points A and B to locate points C and D.

Throw a floating object (e.g., a stick) upstream from points A and C. Observe the object as it floats toward points B and D.

Subtract to find the time it takes for the object to float from start to finish.

Do not attempt to swim across currents that are moving faster than 1.5 meters per second. Equivalents of this speed include—

- Quick-time march rate of 120 counts per minute with one, 30-inch step at each count.
- 5 feet per second.
- 3.5 miles per hour.
- 5.5 kilometers per hour.

Measure River Width

A river's width can be estimated from the width of its symbol on a scaled topographic map. If this is not possible, use the following compass techniques:

Stand at the water line (A).

Shoot an azimuth to a point on the opposite bank (B).

Move upstream or downstream until you are at a point (C) where you can shoot an azimuth 45 degrees larger or smaller than the original azimuth.

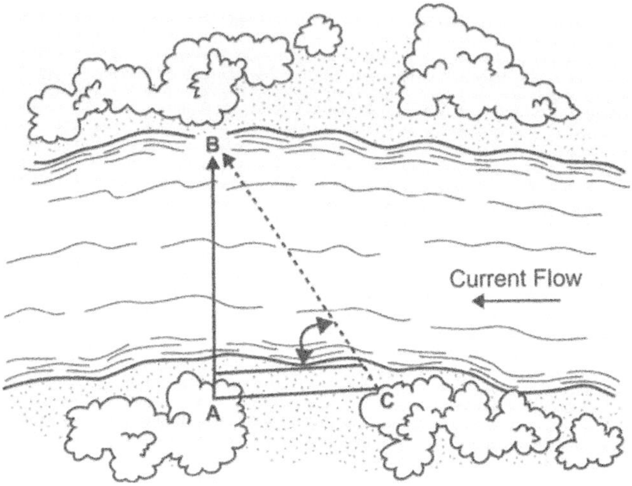

Measure the distance between points A and C. The distance calculated equals the river's width.

Calculate Downstream Drift

A river's current causes personnel and equipment to drift downstream. If personnel and equipment are aimed straight across the river, they will sideslip downstream as they move across the current to the other shore. Therefore, personnel and equipment crossings must compensate for the effects of a river's current and entries are usually made upstream of the desired exit point. Use the formula below to calculate downstream drift.

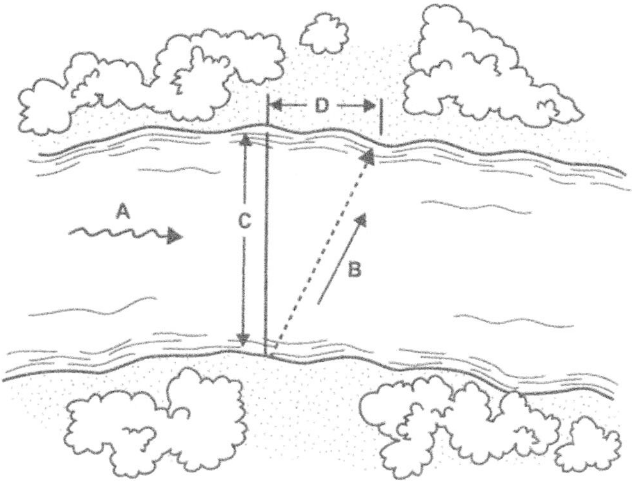

$$\frac{\text{Current Speed (A)}}{\text{Crossing Speed (B)}} \times \text{River Width (C)} = \text{Downstream Drift (D)}$$

> *Note:* The crossing speed for a swimmer across a river may vary but is generally limited to 1 meter per second. All measurements must be in the same unit of measure (e.g., meters, feet).

The Buddy System

Whenever a Marine unit must enter into or operate on the water, a "buddy system" is employed in which every Marine is paired with a swimming partner. The buddy system matches an experienced swimmer with a weak swimmer. The experienced swimmer assists and encourages the weaker swimmer and bolsters confidence during night crossings. If a unit has an odd number of Marines, place the extra person with another pair to form a three-person team.

Water Crossings

Care of Weapons

Marine infantry weapons and munitions are designed to be able to operate after immersion. However, protect your weapons from moisture whenever possible.

A gas-operated weapon can malfunction if water travels down the barrel and enters the gas tube. To protect the gas tube—

Close the weapon's bolt before entering the water.

Seal the muzzle with a condom, balloon, plastic spoon wrapper, or other form of waterproof material.

Tie or melt the protective cover to create a watertight seal.

When the muzzle's protective cover is no longer needed, remove it. Open the bolt and inspect the barrel. If the tactical situation permits, swab excess moisture from the barrel. Test fire automatic weapons, if possible. Field strip and clean weapons as soon as possible. If time does not allow for a complete inspection, rinse inaccessible areas with small amounts of diesel fuel, then dry.

ISOMAT Raft

Construction of an ISOMAT raft is time-consuming. This type of raft should not be employed as part of an attack, but used for logistical purposes (e.g., evacuating stretcher cases, transporting supplies). Use the following steps to build an ISOMAT raft:

Wrap ISOMAT sleeping pads around sturdy sticks.

Use parachute cord and square knots to tie the pads securely in place and to lash stick ends together in a rectangle.

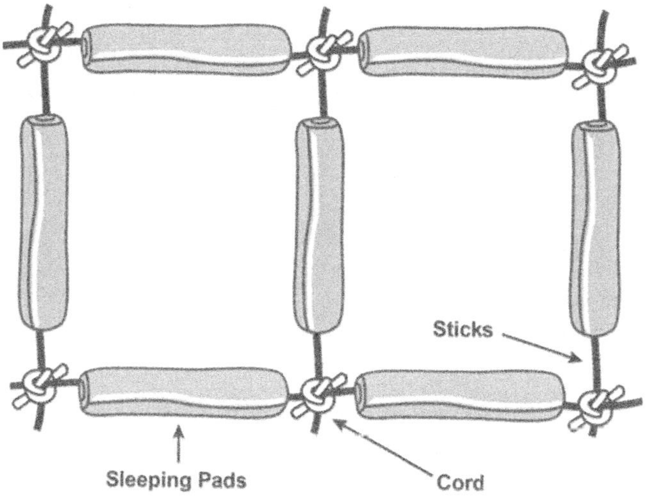

NOTE: The ISOMAT raft pictured can support several hundred pounds. However, the cargo will get wet if not properly waterproofed.

Poncho Raft

A poncho raft can support two Marines and their equipment and is well suited for long crossings. Use the following steps to build a poncho raft:

Inspect two ponchos and ensure they are serviceable.

Lay one poncho flat on the ground, with the hood-side up.

Cinch the hood tightly to form a gooseneck or tie in a knot.

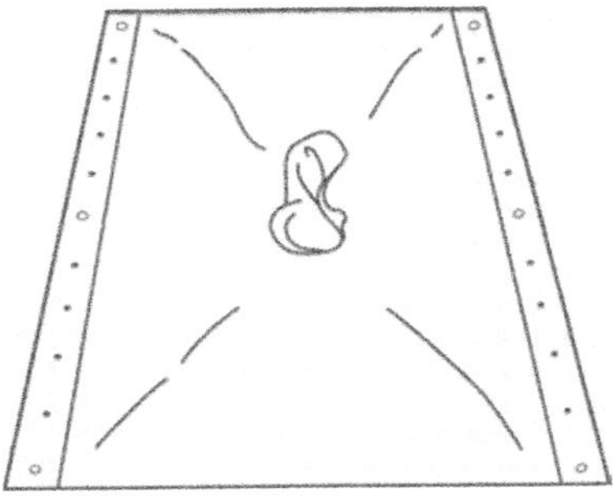

Pad sharp edges of equipment and place the equipment in the center of the poncho.

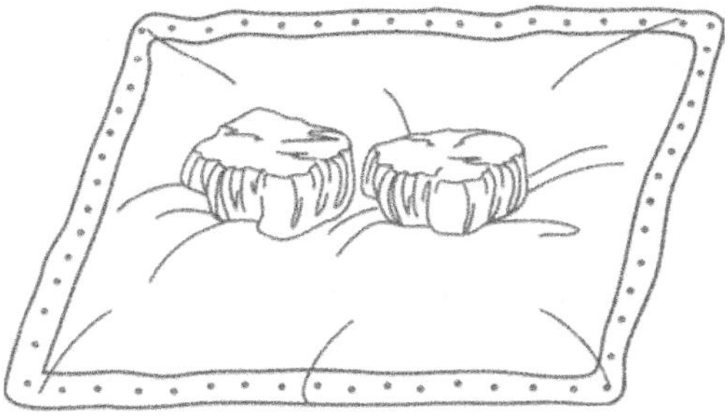

Place the second poncho over the equipment, rubber side up, and hood facing down.

Snap the edges of the two ponchos together.

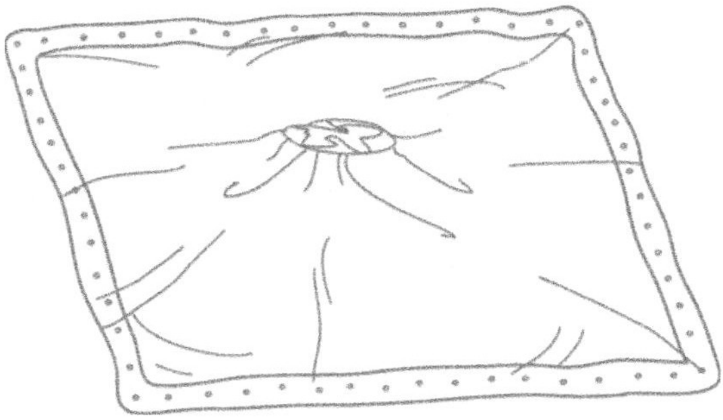

Roll the edges toward the equipment.

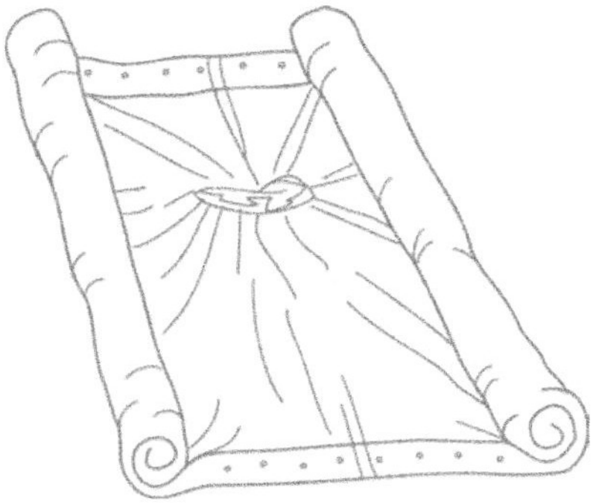

Roll the edges into pigtails and tie them off.

Pull the pigtails together over the top and lash them securely.

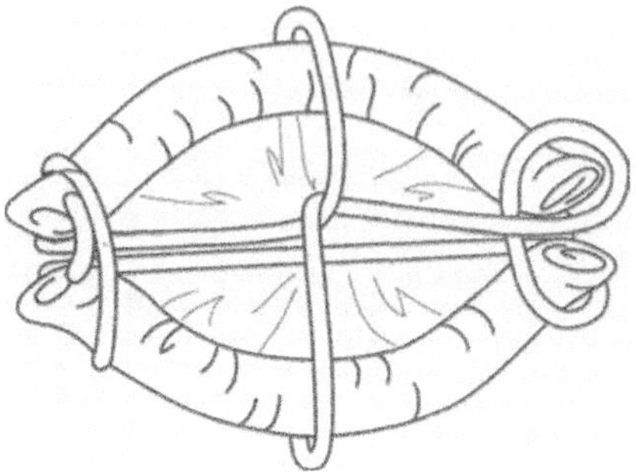

Protect the raft from brush punctures while placing it in the water. Swim across the water obstacle while security elements are covering the far shore.

Construction of a Pack Raft

You will need two waterproofed packs and two M16A2 service rifles. The following steps are required to construct a pack raft:

Place two packs side-by-side with the pack frames on the deck. The tops of the packs are opposite of each other.

Loosen the main compartment straps on both packs.

Insert one rifle on each end between the straps and the packs. The muzzles are opposite of each other. The rifles serve as one means to secure the packs together. Place the front sight post under the top flap.

Tighten the straps so that the rifles and packs are secure.

Take the excess strap on the inner side of each pack and secure it to the opposite pack to better secure the two packs together.

Take the excess straps on the outer sides of the packs and use those straps as safety lashing for the rifles.

Tuck the excess straps and check to make sure the rifles and packs are secure.

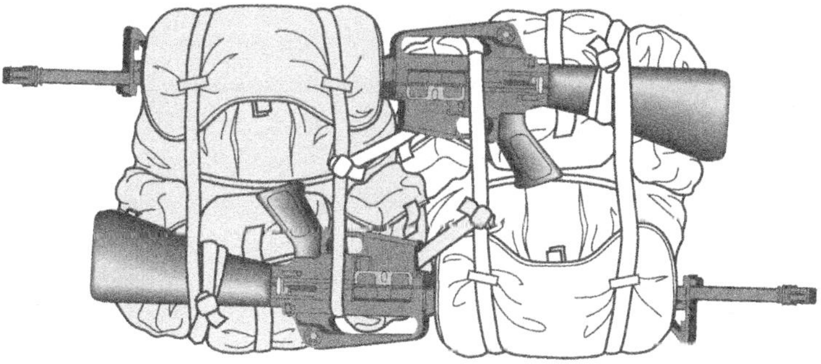

Single-Rope Bridge

A single-rope bridge offers a temporary and quick way to cross small rivers. It also provides extra security while crossing swift waters. At night, it prevents straggling, and guides units precisely from one side of the river to the other side. If crossing a river at night, plan for at least one single-rope bridge.

If your unit is crossing a river with swift currents or water depths above 4 feet, the unit is carrying sufficient rope to span the crossing site, and the tactical situation permits, secure the rope on near and far banks to provide a hand-hold for crossing Marines. This reduces the time required for the entire unit to cross and provides a degree of comfort/confidence for poor swimmers. Use a squad-sized bridge team to construct a single-rope bridge. Station several strong swimmers at the water's edge to help anyone who has trouble crossing.

Nylon rope is normally coiled in 120 foot lengths. It is 0.6 inches in diameter and has a breaking strength of about 3,840 pounds. Over time, a nylon rope can stretch to as much as one-third more than its original length and stretching weakens the rope. If the rope is stretched, discard the rope or use it for light tasks. To prolong the life of a nylon rope, do not step on it or drag it on the ground. Pad the rope in places where it contacts rocks or sharp corners. Do not leave the rope knotted or stretched longer than necessary. Dry rope as soon as possible. Single-rope bridge construction is as follows:

Tie a sling rope around your waist using a square knot and two, separate half hitches. See the appendix for detailed information on knots.

Attach a locking steel carabiner to the sling rope.

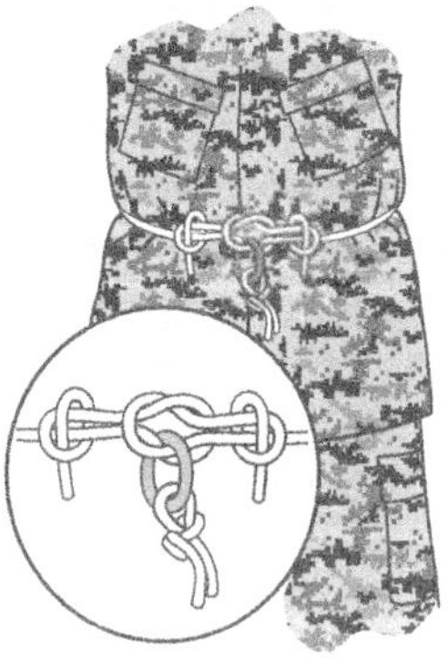

Tie a bowline knot in the running end of the bridge rope and attach it to the carabiner.

Temporarily secure the other end of the rope to a tree on the near shore.

Enter and cross the water.

> NOTE: Carry only your weapon and ammunition.

Exit the water on the opposite shore.

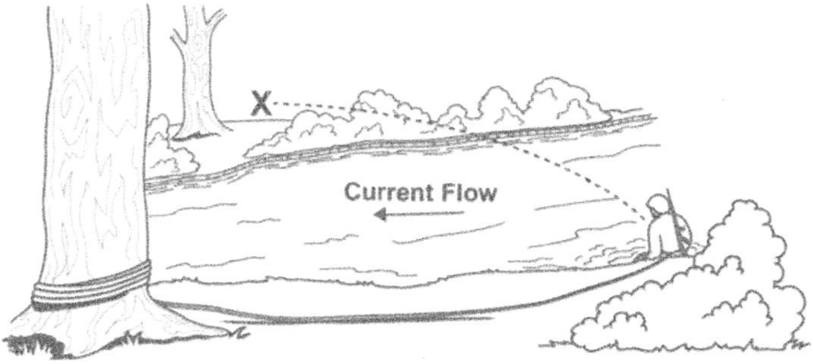

Prepare your weapon for use. Unhook the bridge rope from the carabiner at your waist, and tie the bridge rope to a sturdy tree using a round turn and two half hitches.

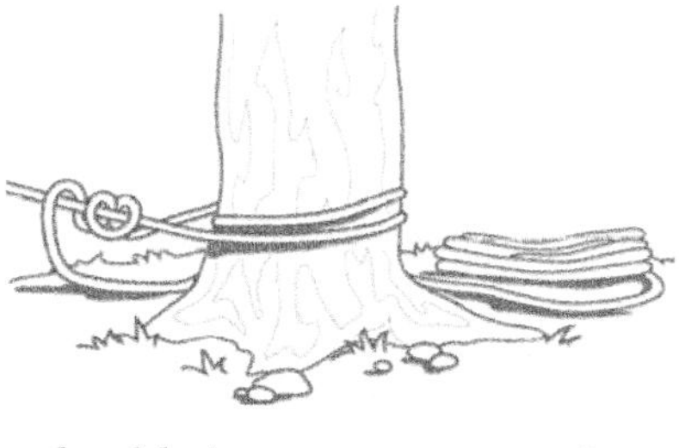

Conduct a box reconnaissance of the opposite shore.

On the near shore, have another Marine prepare to tighten the rope. That Marine should place a transport tightening system in the bridge rope by tying a double butterfly knot and placing two carabiners in the butterfly. See the appendix for detailed information on knots.

The Marine should pass the running end of the bridge rope around the downstream side of the near shore anchor point and through the two carabiners.

Pull the butterfly knot approximately one-third of the distance across the river.

Secure the bridge rope to an anchor point using a round turn and two half-hitches.

On the near shore, the Marine helping you should pull the slack out of the bridge rope until the butterfly knot is back on the near side. The bridge rope is then tied off against itself using two half hitches with a quick release in the last half hitch.

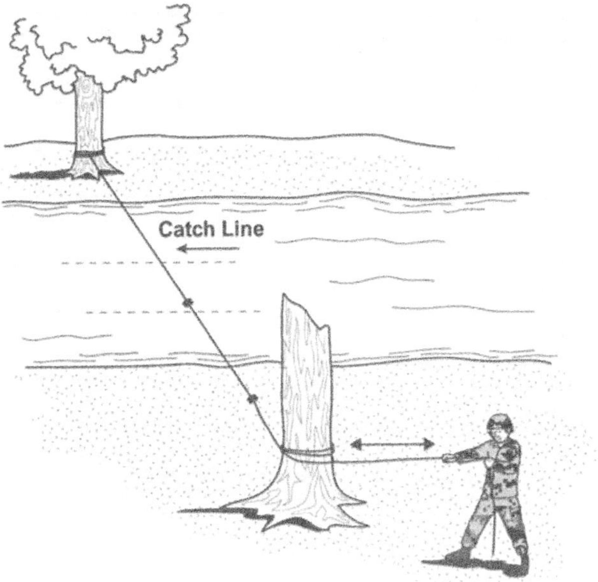

NOTE: The single-rope bridge must be as tight as possible so it will not sag when used.

If you lose your footing and fall into the water, swim with the current to the closest shore. Swimming against the current is dangerous and quickly causes fatigue.

High and Dry Crossings

If the single-rope bridge is high enough, suspend yourself below the single-rope bridge and above the water. Use the following steps to suspend yourself from a single-rope bridge and then pull yourself across the water:

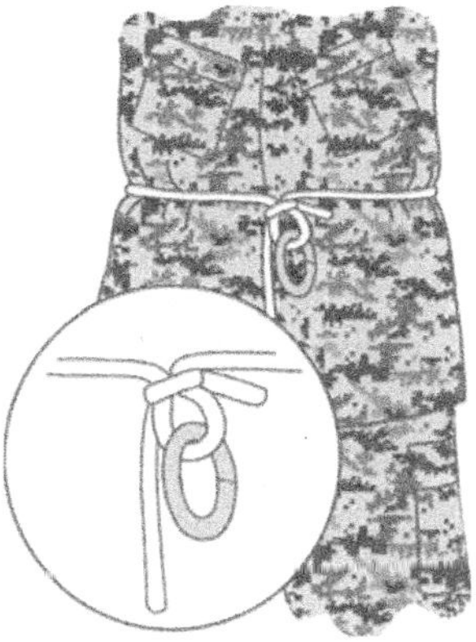

Tie a sling rope around your waist using a bowline. Ensure that the knot is tight.

Attach a carabiner through the bowline's loop. The carabiner's gate faces up.

Secure your helmet chin strap.

Face the single-rope bridge with your left shoulder toward the far shore.

Grasp the bridge rope in both hands.

Swing your body beneath the single-rope bridge with your head toward the far shore. Cross your ankles above the bridge rope.

Arch back until the carabiner contacts the bridge rope. Connect the carabiner to the bridge rope. Allow the carabiner to bear your body's weight.

Pull yourself across the single-rope bridge, hand over hand, to the far shore.

Swift Current Crossings

A single-rope bridge prevents being knocked down and swept away by a swift current. Use the following steps to move through a swift current:

Tie one end of a sling rope around your waist using a bowline.

Tie the running end of the sling rope in another bowline, and attach a carabiner to the bowline's loop.

Step up to the bridge. Face upstream.

Hook the carabiner to the single-rope bridge.

Walk sideways into the river while grasping the bridge rope in both hands.

Use the single-rope bridge for balance and remain standing, if possible.

Continue to move sideways through the river to the far shore.

Slow Current Crossings

If you face little or no current, it is not necessary to hook up to a bridge rope with a carabiner. Lie on your back in the water beneath the single-rope bridge. Support your body weight with your waterproof pack. Use the bridge rope and pull yourself hand over hand across the river.

Removal

If you are the last Marine waiting to cross, pull on the standing end of the rope to release the knot, then tie the rope around your waist using a bowline. The Marines on the far shore will pull you through the water.

Appendix
Knot Tying

Square Knot

The square knot is used to secure two ropes of equal diameter together so they form one continuous rope that will not slip.

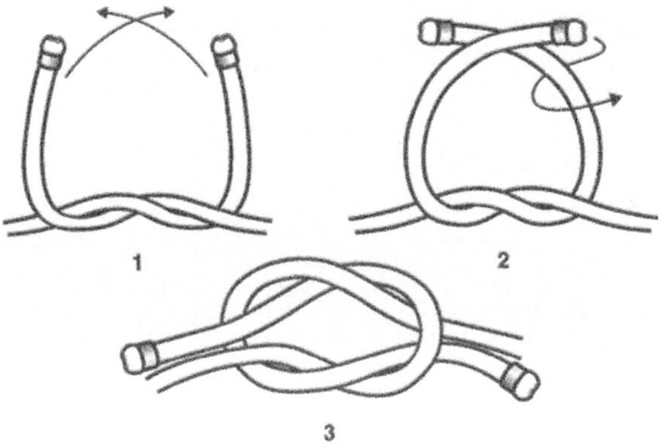

Bowline

The bowline forms a loop that will not tighten or slip under strain. It is easily untied.

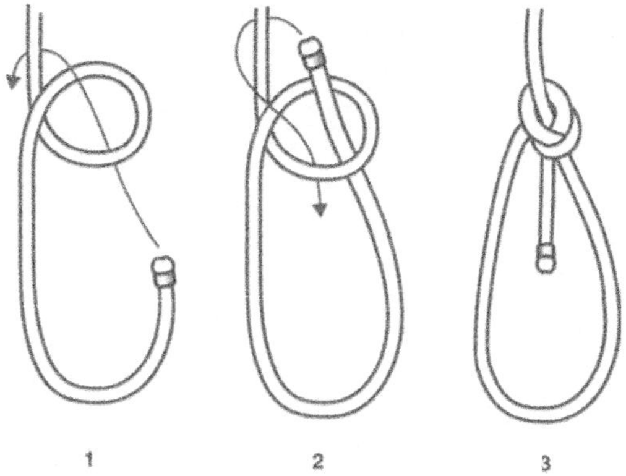

Hitches

Hitches are used to dress (prepare) knots and secure loose ends.

Half Hitch

The half hitch is used to tie a rope to a tree or to a larger rope. It will hold against a steady pull, but is not a secure hitch. It is frequently used to secure the free end of a rope.

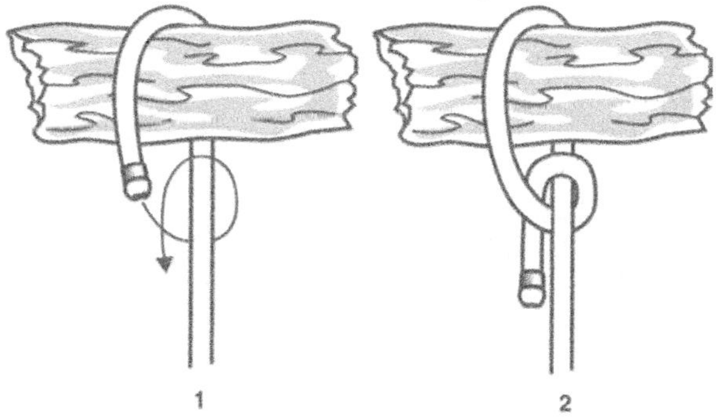

Two Half Hitches

Two half hitches can be used to secure the running end of a rope.

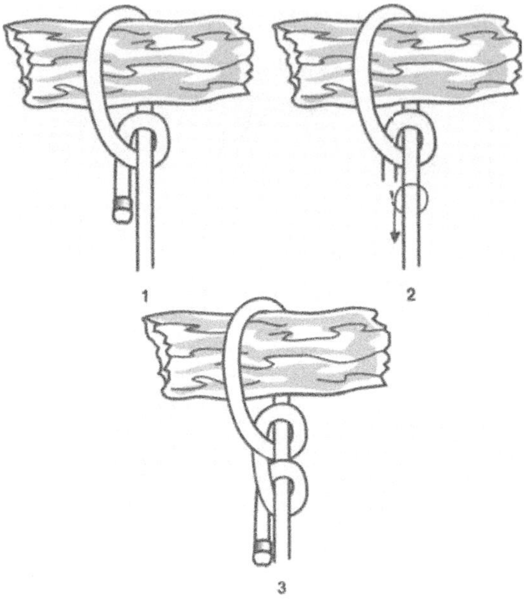

Round Turn and Two Half Hitches

A round turn and two half hitches can be used to fasten a rope to a tree. This hitch does not jam.

Butterfly Knot

The butterfly knot is used for anchor lines. Tied properly, it will not tighten on itself to the point that it cannot be easily untied. The butterfly knot can be used to pull a rope bridge taut. This knot can be used to tighten a fixed rope when mechanical means are not available. It will not jam if a stick is placed between the two upper loops.

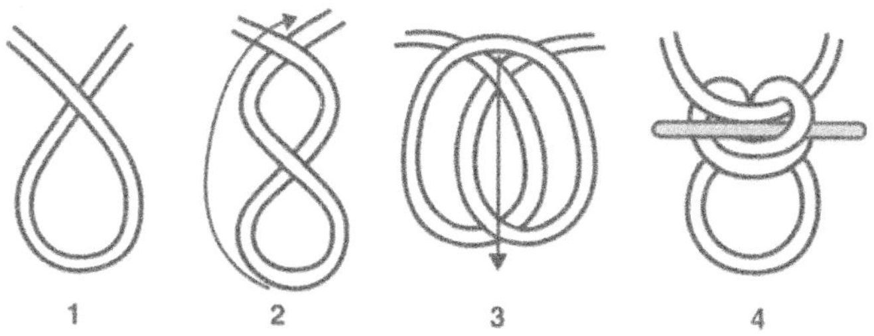

TABLE OF CONTENTS

Chapter 1. The M9 Service Pistol

Chapter 2. Weapons Handling

Chapter 3. Fundamentals of Pistol Marksmanship

Chapter 4. Pistol Firing Positions and Grip

Chapter 5. Use of Cover and Concealment

Chapter 6. Presentation of the M9 Service Pistol

Chapter 7. Pistol Engagement Techniques

CHAPTER 1
THE M9 SERVICE PISTOL

The Marine Corps' uses the M9 service pistol as a defensive weapon. It is a semiautomatic, magazine-fed, recoil-operated, double-action pistol. The M9 service pistol's magazine holds 15 rounds and is chambered with a 9-mm ball (NATO M882 round). The M9 service pistol has a maximum effective range of 50 meters (54.7 yards).

> *Note:* The procedures in this chapter are written for right-handed Marines; left-handed Marines reverse instructions as necessary.

Functional Capabilities

When the M9 service pistol's safety lever is moved to the firing position, it can be fired in either a single-action or double-action mode. It is designed to fire one round each time the trigger is pulled. When the last round is fired, the slide automatically locks to the rear.

Single-Action Mode

The single-action mode allows the pistol to be fired when the hammer is cocked; single action requires the hammer to be cocked to the rear before the trigger is pulled. The hammer can be manually cocked or mechanically cocked. The hammer is mechanically cocked after the first shot is fired. See figure 1-1.

Figure 1-1. Single-Action Mode.

Double-Action Mode

The double-action mode causes the hammer to move to the rear as the trigger is being pulled. See figure 1-2.

Figure 1-2. Double-Action Mode.

Nomenclature

See figure 1-3 on page 1-2.

Major Components

The M9's major components consist of the slide assembly, barrel assembly, and receiver. See figure 1-4 on page 1-3.

Slide Assembly

The slide assembly houses the firing pin, firing pin block, striker, extractor, and loaded chamber indicator, and it cocks the hammer during recoil.

Extractor

The extractor pulls the brass from the chamber after the round is fired.

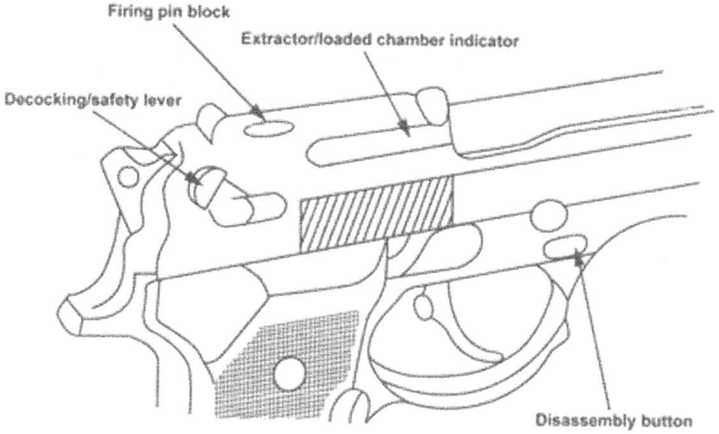

Right Side View

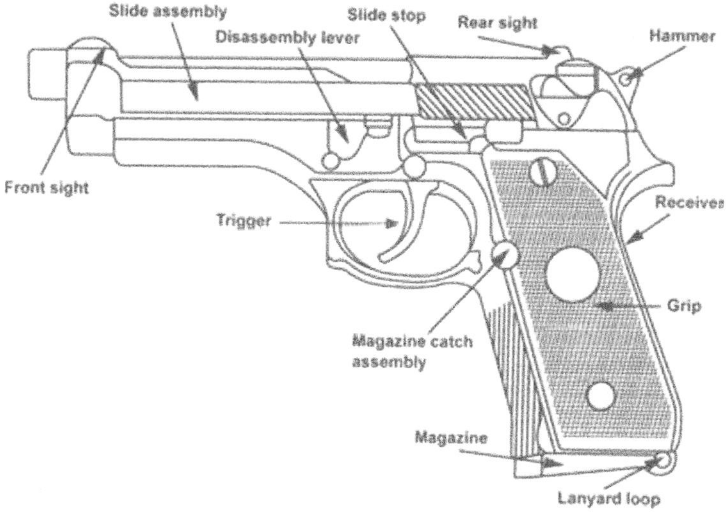

Left Side View

Figure 1-3. M9 Service Pistol.

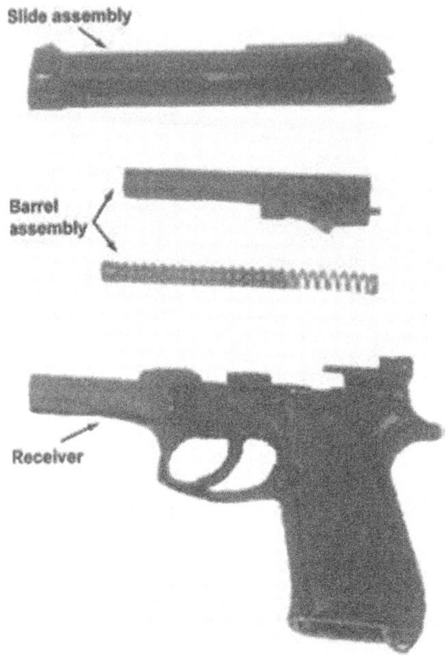

Figure 1-4. Major Components.

Loaded Chamber Indicator

When a round is in the chamber, the upper surface of the loaded chamber indicator protrudes from the right side of the slide. This protrusion can be felt with the finger, verifying that there is a round in the chamber.

Barrel Assembly

The barrel assembly houses the round for firing, directs the projectile, and locks the barrel in position during firing.

Receiver

The receiver supports the major components, controls the functioning of the pistol, and holds the magazine in place. The front and back straps of the grip are grooved vertically to ensure that the hand does not slip while firing. The receiver consists of the disassembly button, slide stop, and magazine catch assembly.

Disassembly Button

The disassembly button permits quick disassembly of the pistol.

Slide Stop

The slide stop holds the slide to the rear after the last round is fired. It can also be manually operated to lock the slide to the rear or release the slide.

Magazine Catch Assembly (Magazine Release Button)

The magazine catch assembly secures the magazine in place when loading, and it releases the magazine from the pistol when unloading. The magazine catch assembly is designed for both right- and left-handed Marines.

> *Note:* Reversal of the magazine catch assembly for left-handed Marines can be performed by a qualified armorer.

Safety Features

The safety features of the M9 service pistol include the decocking/safety lever, firing pin block, and half-cock notch.

Decocking/Safety Lever

The decocking/safety lever, commonly referred to as the safety, permits safe operation of the pistol by both right- and left-handed Marines. As the safety is moved to the safe (down) position, the firing pin striker moves out of alignment with the firing pin. This movement prevents the pistol from firing as the hammer moves forward.

> *Note:* In the fire (up) position, a red dot is visible, indicating that the pistol is ready to fire.

Firing Pin Block

The firing pin block rests in the firing pin notch and prevents movement of the firing pin until the trigger is pulled. As the trigger is pulled, the firing pin block moves up and out of the firing pin notch. This movement allows a round to be fired when the hammer strikes the firing pin.

Half-Cock Notch

The half-cock notch stops the forward movement of the hammer during a mechanical failure.

Cycle of Operation

There are eight steps in the cycle of operation for the M9 service pistol.

Firing

Once the safety is off and the trigger is pulled to the rear, the hammer falls on the firing pin, which strikes the primer and ignites the round. See figure 1-5.

Figure 1-5. Firing.

Unlocking

As the slide assembly moves to the rear, the locking block rotates out of the notches in the slide. See figure 1-6.

Figure 1-6. Unlocking.

Extracting

As the slide moves rearward, the extractor withdraws the cartridge case out of the chamber. See figure 1-7.

Figure 1-7. Extracting.

Ejecting

As the face of the slide passes over the ejector, the case strikes the ejector and it is knocked upward and outward through the ejection port. See figure 1-8.

Figure 1-8. Ejecting.

Cocking

As the slide moves rearward, the hammer is pushed back, allowing the sear to engage the hammer hooks, cock the hammer to the rear, and place the pistol in the single-action mode. See figure 1-9.

Figure 1-9. Cocking.

Feeding

The slide starts forward, pushed by the recoil spring. The face of the slide makes contact with the cartridge at the top of the magazine, stripping it from the magazine and pushing it toward the chamber. See figure 1-10.

Figure 1-10. Feeding.

Chambering

As the slide continues forward, it pushes the cartridge into the chamber. See figure 1-11.

Figure 1-11. Chambering.

Locking

As the slide assembly continues to move forward, the locking block lugs move into the locking block recesses on the right and left sides of the slide. See figure 1-12 on page 1-6.

Figure 1-12. Locking.

Ammunition

The only ammunition authorized for the M9 service pistol is the NATO M882 9-mm ball. Dummy ammunition can be used during training (a dummy round has a hole drilled in its side and contains no primer). See figure 1-13.

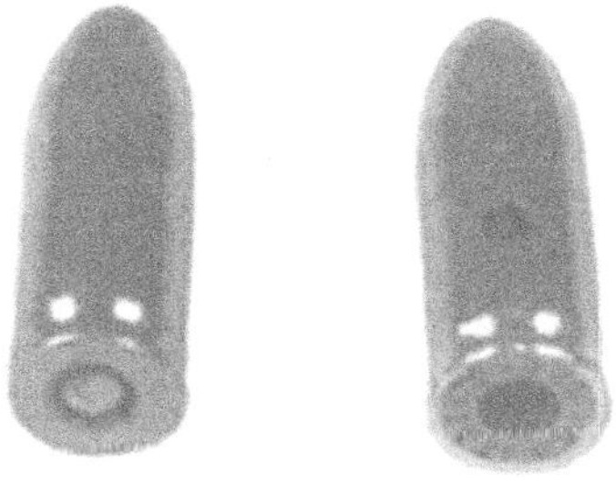

Figure 1-13. M9 Service Pistol Ammunition.

Do not open ammunition containers until the ammunition is to be used. Ammunition must be maintained in a high state of readiness. To care for ammunition—

- Keep ammunition dry and clean. If ammunition gets wet or dirty, wipe it off with a clean dry cloth.
- Wipe off light corrosion as soon as it is discovered. Never use ammunition that is heavily corroded, dented, or has the projectile pushed in.
- Do not expose ammunition to direct sunlight for long periods of time.
- Do not oil or grease ammunition. Dust or other abrasives can collect on greasy ammunition and may cause damage to the operating parts of the pistol. Oiled cartridges also produce excessive chamber pressure.

Wearing of the M9 Service Pistol's Gear

The proper placement of pistol gear helps ensure safety and aids the Marine in effectively handling and employing the pistol.

M12 Holster

The M12 holster consists of the holster, removable holster flap, and metal retaining clip. To check for proper placement of the holster, allow the right arm to hang freely. The holster should be slightly in front of the arm to permit easy access to the pistol upon presentation from the holster. See figure 1-14.

Note: In most cases, the holster is issued with the holster flap installed for a right-handed Marine. To convert the holster for a left-handed Marine, remove the metal retaining clip and install the clip on the opposite side of the holster.

Figure 1-14. The M12 Holster and Ammunition Pocket.

M1 Ammunition Pocket

The ammunition pocket attaches to the cartridge belt on the side opposite the holster in a position that best permits ready access for reloads. The magazine is stored in the ammunition pocket with the rounds down and pointed inboard.

M7 Shoulder Holster

The M7 shoulder holster consists of a holster with a thumb snap closure, shoulder strap, chest strap, and a belt retaining loop. The holster is positioned on the left side of the chest to provide easy and quick access with the right hand. See figure 1-15. The M7 holster comes fully assembled and has adjustable straps to accommodate each Marine's body size. The holster is available for right-handed Marines only, therefore, a left-handed Marine has to withdraw the pistol from the holster with the right hand and then transfer the pistol to the left hand before firing. (See chap. 8 for transferring the pistol from one hand to the other.) When the shoulder holster is worn properly—

Figure 1-15. M7 Shoulder Holster.

- The shoulder strap lays flat across the left shoulder with the shoulder pad directly on top of the shoulder.
- The chest strap attaches to the D-ring at the top of the holster and runs diagonally across the chest, underneath the right arm, and around the back where it attaches to the end of the shoulder strap.
- The belt retaining loop is at the bottom of the holster and attaches to the belt to stabilize the holster's position.
- The ammunition pocket attaches to the chest strap directly underneath the right arm.

Assault Holster

Some Marines (i.e., Marine security force, direct action platoon, and military police) are required to carry the assault holster. This holster has a retention strap that fastens over the top of the holster to retain the pistol. This type of holster generally has a thumb break on the retention strap that is disengaged to access the pistol. See figure 1-16.

Figure 1-16. Assault Holster.

Concealed Pistol Holster

Some Marines are required to carry a concealed pistol as part of their official duties. The primary consideration for placement of a concealed pistol holster is to ensure the pistol cannot be seen; therefore, the Marine must consider the type of clothing to be worn. A secondary consideration is to place the holster so the pistol can be presented easily. Typically, the best position for a concealed holster is just behind the strong side hip. This position best conceals the pistol while allowing it to be presented quickly. Another placement choice is in a shoulder holster, placing the pistol just under the weak side arm. See figures 1-17 through 1-20 on pages 1-8 and 1-9.

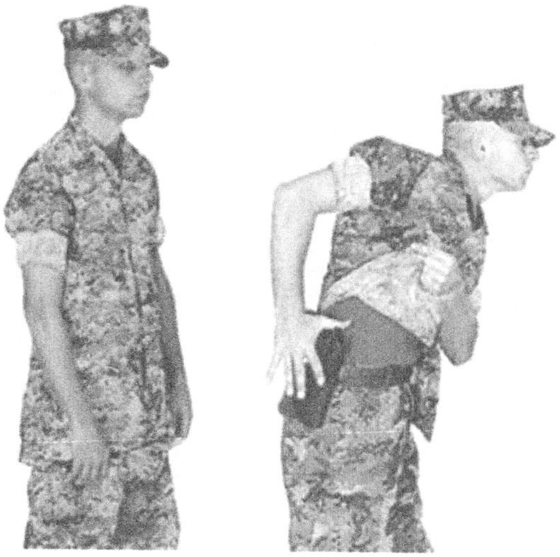

Figure 1-17. Wearing of the Concealed Pistol Holster—Utilities.

Figure 1-18. Wearing of the Concealed Pistol—Sweater.

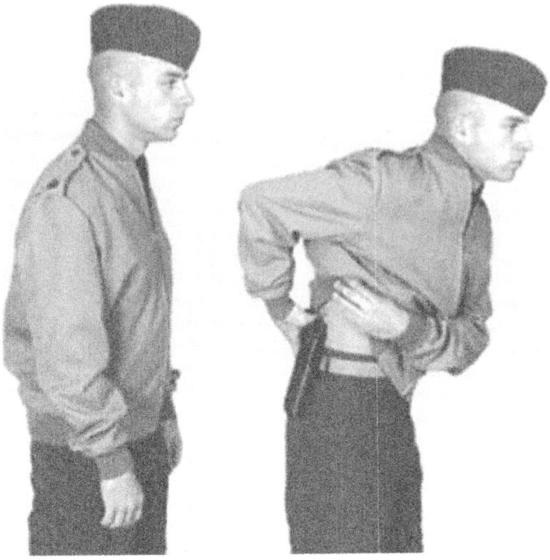

Figure 1-19. Wearing of the Concealed Pistol Holster—Jacket.

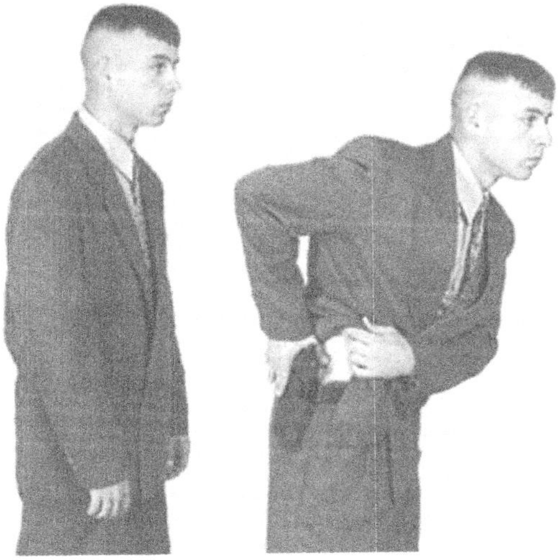

Figure 1-20. Wearing of the Concealed Pistol Holster—Civilian Attire.

Lanyard

The lanyard aids in pistol retention. It is issued in three sizes and is adjustable. The lanyard consists of a fabric cord, two cylindrical slip rings, and a metal clip that attaches the lanyard to the pistol's lanyard loop. See figure 1-21.

Figure 1-21. M9 Service Pistol (With Lanyard).

To don the lanyard—

- Adjust the slip rings so they are positioned flush with the lanyard's base (clip end).
- Place the right arm through the loop and place the loop over the head, resting on the left shoulder.
- Attach metal clip to the pistol's lanyard loop.
- Place the pistol in the holster.
- Use the left hand to hold the bottom slip ring against the base of the lanyard. Use the right hand to slide the top slip ring upward to position the loop of the lanyard under the arm. The lanyard should fit snugly against the body, but not restrict the Marine's movements.
- Tuck any excess cord behind the holster.
- Ensure that the lanyard is adjusted properly by removing the pistol from the holster and fully extending the right arm. The lanyard should be taut. Adjust as necessary.

Note: Re-adjust the lanyard if any equipment changes are made (e.g., flak jacket).

Firing the M9 Service Pistol While Wearing Gloves

Not all combat engagements occur during ideal weather conditions. During cold weather, the Marine may find it necessary to wear gloves to protect the fingers from frostbite and help prevent stiffening of the hands. Gloves may also be worn in mission-oriented protective posture conditions. Gloves provide protection to the hands, however, they also may interfere with the Marine's ability to engage targets effectively. The added bulk of the gloves may affect the Marine's ability to manipulate the safety, magazine release button, magazine, hammer, and slide stop/release. For example, the Marine may need to exert more pressure to engage the magazine release or slide release buttons to compensate for the thickness of the gloves.

The principles of target engagement (see chap. 7) do not change while wearing gloves, however, the specific ability to manipulate and control the trigger is greatly affected by the thickness of the gloves around the fingers. Wearing gloves reduce the Marine's ability to feel, which makes it difficult to apply trigger control when firing. The Marine may find that more pressure than normal must be applied with the trigger finger just to establish initial contact with the trigger. Once the Marine can "feel" the trigger through the gloves, then the pressure required to fire a shot can be applied. This action may increase the chances of firing the pistol prematurely due to excessive pressure on the trigger. Dry firing while wearing gloves allows the Marine to learn how to apply trigger control consistently and determines how much pressure is needed to effectively fire a shot.

While wearing gloves, the Marine may find it difficult to fire the pistol in the double-action mode due to the position of the trigger and the limited amount of space between the trigger and the trigger guard. Therefore, if the situation permits, the Marine may wish to thumbcock the pistol to fire in single-action mode. In single-action mode, there is more space between the trigger guard and the trigger, making it easier to position the finger on the trigger. However, the Marine's ability to thumbcock the pistol may also be hindered by the gloves' bulk. Therefore, to thumbcock the pistol while wearing gloves, the Marine may perform one of the following methods:

- For method one, loop a section of 550 cord (approximately 2 inches) through the loop located on the top rear portion of the hammer. The length of the cord should not interfere with the pistol's cycle of operation or with the Marine's ability to establish sight alignment. Once the cord is attached to the hammer, take the pistol off safe and pull downward on the cord to cock the hammer. See figure 1-22.

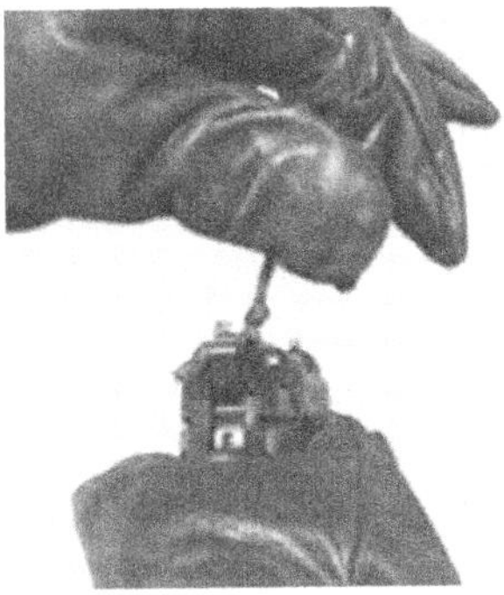

Figure 1-22. Cocking the Pistol with 550 Cord.

- For method two, take the pistol off safe, rotate the pistol inboard, and place the top of the hammer against a secure surface (e.g., cartridge belt, table top, heel of boot). Apply pressure on the pistol to keep the hammer in place and push downward on the pistol in one continuous motion to cock the hammer. See figure 1-23.

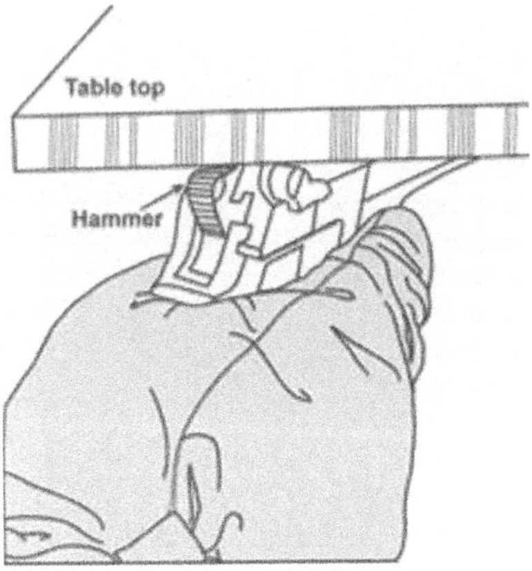

Figure 1-23. Cocking the Pistol on a Secure Surface.

Preventive Maintenance

If the M9 service pistol is to be effective, it must be maintained in a state of operational readiness at all times; therefore, maintenance of the M9 service pistol is a continuous effort. A clean, properly lubricated, well-maintained pistol will fire when needed.

Pistol Disassembly

Before disassembling the M9 service pistol, ensure that the pistol is in Condition 4. The pistol is in Condition 4 when the magazine is removed, the chamber is empty, the slide is forward, and the safety is on. To disassemble the pistol, perform the following steps in sequence:

- Hold the pistol in the right hand with the muzzle slightly elevated. Reach over the slide with the left hand and place the left index finger on the disassembly button and the left thumb on the disassembly lever. Press the disassembly button and hold it in place while rotating the disassembly lever downward until it stops.

Note: A left-handed Marine places the right thumb on the disassembly button and the right index finger on the disassembly lever.

- Pull the slide and barrel assembly forward and remove it while wrapping the fingers around the slide to hold the recoil spring and recoil spring guide in place.
- Turn the slide assembly over in the left hand until the recoil spring and recoil spring guide face up. Place the right thumb on the end of the recoil spring guide next to the locking block and compress the recoil spring and spring guide while lifting and removing them from the slide and barrel assembly. See figure 1-24. Allow the recoil spring to decompress slowly.

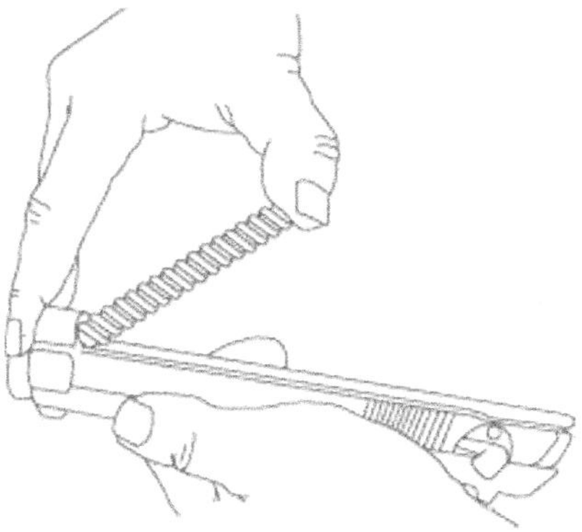

Figure 1-24. Removing the Recoil Spring and Recoil Spring Guide.

- Separate the recoil spring from the recoil spring guide.
- Push in on the locking block plunger with the right index finger while pushing the barrel forward slightly. Lift and remove the locking block and barrel assembly from the slide.

A Marine is not authorized to disassemble the pistol any further than the preceding steps. Any further disassembly must be performed by ordnance personnel. See figure 1-25.

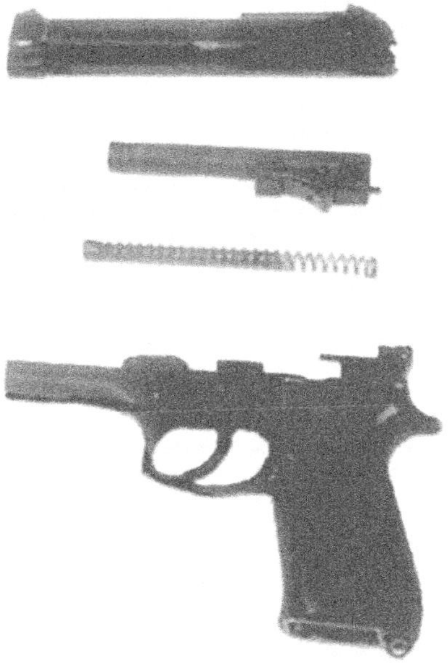

Figure 1-25. Disassembled M9 Service Pistol.

Disassembly of the Magazine

To disassembly the magazine, perform the following steps:

- Grip the magazine firmly in the left hand with the floorplate up and the thumb resting against the flat end of the floorplate.
- Release the floorplate by pushing down (with a blunt object; e.g., an ink pen) on the floorplate retainer stud in the center of the floorplate. At the same time, slide the floorplate a short distance forward with the thumb.
- Maintain the magazine spring pressure with the thumb and remove the floorplate from the magazine.
- Remove the floorplate retainer and magazine spring and follower from the magazine tube.

See figure 1-26.

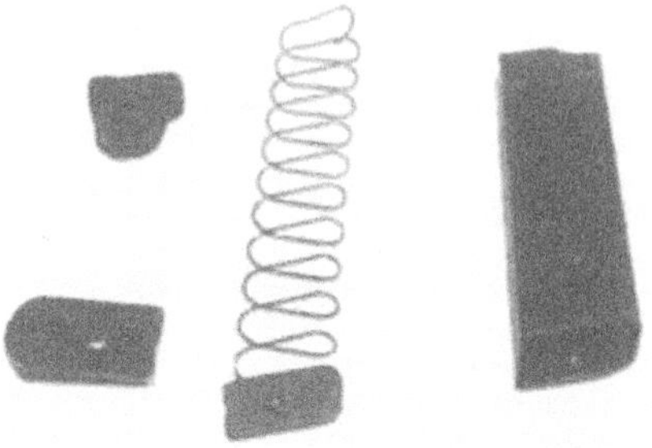

Figure 1-26. Disassembled Magazine.

Inspection of the Pistol

Once the pistol has been disassembled, it must be thoroughly inspected to ensure it is in a serviceable condition. Pistol inspection is continuous during the pistol's cleaning and reassembly:

Slide Assembly

- Check for free movement of the safety. Ensure the rear sight is secure.

Barrel Assembly

- Inspect the bore and chamber for pitting or obstructions.
- Check the locking block plunger for free movement of the locking block.
- Inspect the locking lugs for cracks and burrs.

Recoil Spring and Recoil Spring Guide

- Check the recoil spring for damage.
- Check that it is not bent.
- Check the recoil spring guide for straightness and smoothness.
- Check to be sure it is free of cracks and burrs.

Receiver Assembly

- Check for bends, chips, and cracks.
- Check for free movement of the slide stop and magazine catch assembly.
- Check the guide rails for excessive wear, burrs, cracks, or chips.

Magazine Assembly

- Check the spring and follower for damage.
- Ensure the lips of the magazine are not excessively bent and are free of cracks and burrs. The magazine tube should not be bent or dirty.

Cleaning and Lubricating the Pistol

Only authorized cleaning materials should be used to clean and lubricate the pistol. If these items are not issued with the pistol, they may be obtained from the armory. The following procedures are used to clean and lubricate the pistol—

Slide Assembly

- Clean the slide assembly with a cloth. A general purpose brush and cleaning lubricant protectant (CLP) can also assist in the removal of excess dirt and carbon buildup.
- Ensure the safety, bolt face, slide guides, and extractor are free of dirt and residue.
- Wipe dry with a cloth and apply a light coat of CLP.

Barrel Assembly

- Insert a bore brush with CLP into the chamber end of the barrel, ensuring that it completely clears the muzzle before it is pulled back through the bore.

Caution

Insert the bore brush through the chamber to prevent damage to the crown of the barrel.

- Repeat several times to loosen carbon deposits.
- Dry the barrel by pushing a swab through the bore.
- Repeat until a clean swab can be observed.
- Clean the locking block with a general purpose brush.
- Use the barrel brush to apply a light coat of CLP to the bore and chamber area and lubricate the exterior surfaces of the barrel and locking block.

Recoil Spring and Recoil Spring Guide

- Clean the recoil spring and recoil spring guide using CLP and a general purpose brush or cloth.
- Apply a light coat of CLP after wiping the recoil spring and recoil spring guide clean.

Receiver

- Wipe the receiver assembly clean with a cloth.
- Use a general purpose brush for areas that are hard to reach, paying special attention to the

disassembly lever, trigger, slide stop, hammer, and magazine release button.

- Apply a light coat of CLP.

Caution

Do not allow the hammer to fall with full force by pulling the trigger when the slide is removed. This can damage the receiver and hammer.

Magazine

- Clean the magazine tube and follower with CLP and a general purpose brush.
- Wipe the magazine spring, floorplate retainer, and floorplate clean with a cloth.
- Apply a light coat of CLP.

Pistol Re-assembly

After the M9 service pistol has been cleaned and lubricated, it must be properly re-assembled to ensure its serviceability. To re-assemble the M9 service pistol—

- Use the left hand to grasp the slide with the bottom facing up and the muzzle pointing toward the body. Use the right hand to grasp the barrel assembly with the locking block facing up. Use the index finger to push in the locking block plunger while placing the thumb on the base of the locking block.
- Insert the muzzle of the barrel assembly into the forward open end of the slide. At the same time, lower the rear of the barrel assembly by slightly moving the barrel downward. The locking block should fall into the notches of the slide assembly.
- Slip the recoil spring guide into the recoil spring.
- Insert the end of the recoil spring and recoil spring guide into the slide recoil spring housing. At the same time, compress the recoil spring and lower the spring guide until it is fully seated onto the locking block cutaway.
- Use the left hand to grasp the slide and barrel assembly, sights up, and wrap the fingers around the slide assembly to hold the recoil spring and guide in place. Align the slide assembly guide rails onto the receiver assembly guide rails.
- Push the slide rearward while pushing up on the slide stop with the thumb. Lock the slide to the rear while maintaining upward pressure on the slide stop. Rotate the disassembly lever upward. Listen for a click, an audible click indicates a positive lock.

Pistol Magazine Re-assembly

To re-assemble the magazine—

- Grip the magazine firmly in the left hand with the floorplate end up and the counting holes facing the Marine. Insert the follower into the magazine so the flat end of the follower is against the flat end of the magazine.
- Ensure the floorplate retainer is attached to the first curve of the bottom coil.
- Hold the spring upright with the right hand and insert the spring into the magazine tube so that the flat end of the floorplate retainer is against the flat end of the magazine.
- Push the magazine spring and floorplate retainer down with the right hand and hold it in place with the thumb of the left hand. Use the right hand to slide the floorplate over the side walls of the magazine until fully seated, which is indicated by an audible click.

Safety/Function Check

A safety/function check is performed after reassembling the M9 service pistol. Perform the following steps to ensure the pistol is operational:

- Ensure there is no ammunition in the chamber of the pistol.
- Ensure that the safety is in the safe position, then depress the slide stop, allowing the slide to return fully forward. At the same time, the hammer should fall to the full forward position.

- Pull and release the trigger. The firing pin block should move up and down but the hammer should not move.
- Place the safety in the fire position.
- Pull the trigger to check the double action. The hammer should cock and fall.
- Pull the trigger again and hold it to the rear. Use the fingers and thumb of the left hand to grasp the serrated sides of the slide just forward of the safety. Pull the slide to its rearmost position and release it while holding the trigger to the rear. Release the trigger, a click should be heard and the hammer should not fall.
- Pull the trigger to check the single action. The hammer should fall. Place the safety in the safe position.

If the safety/function check does not indicate an operational pistol, the Marine takes the pistol to organizational maintenance or the next authorized repair level.

User Serviceability Inspection

The Marine is responsible for performing a user serviceability inspection on the pistol prior to live fire. The user serviceability inspection ensures the pistol is in an acceptable operating condition. This inspection is not intended to replace the detailed pistol components inspection following disassembly or the limited technical inspection or pre-fire inspection conducted by a qualified armorer. To conduct a user serviceability inspection on the pistol, perform the following steps:

- Ensure the magazine release button is on the left side of the pistol for a right-handed Marine, the right side of the pistol for a left-handed Marine.
- Ensure the magazine seats into the magazine well when it is inserted and that it cannot be pulled out.
- Ensure the slide stays locked to the rear when the slide is pulled rearward with an empty magazine in the pistol.
- Ensure the magazine falls out freely when the magazine release button is depressed.
- Repeat the preceding four steps with the second magazine.
- Ensure the slide is locked to the rear before lubricating the spring guide, the top of the barrel just forward of the front sight, and the guide rails of the slide assembly behind the safety. With the muzzle pointed downward, work the slide several times and release.
- Inspect the pistol's external parts visually to ensure that there are no cracks or excessive wear.
- Perform a safety/function check of the pistol.

Pistol Maintenance in Adverse Conditions

Combat situations can place Marines in a variety of adverse conditions. Therefore, the M9 service pistol must be maintained properly to ensure its continued operation.

Extreme Cold

In extreme cold conditions, ensure that the following maintenance is performed:

- Clean and lubricate the pistol inside at room temperature, if possible.
- Apply a light coat of lubricant, arctic weather (LAW) to all functional parts.
- Always keep the pistol dry.
- Keep the pistol covered when moving from a warm to a cold area. This permit gradual cooling of the pistol and prevents freezing.
- Ensure that a hot pistol is not placed in snow or on ice.
- Keep snow out of the bore of the barrel. If snow should enter the bore, use a swab and cleaning rod to clean the bore before firing.

Hot, Wet Climates

Maintenance in hot, wet climates must be performed more frequently:

- Inspect hidden surfaces for corrosion. If corrosion is found, clean and lubricate.
- Remove handprints with a clean cloth in order to prevent corrosion.
- Dry the pistol with a cloth and lubricate it with CLP.
- Check ammunition and magazines frequently for corrosion. Disassemble and clean the magazines with CLP and wipe dry with a clean cloth. If necessary, clean ammunition with a dry cloth.

Hot, Dry Climates

In a hot, dry climate, dust and sand can cause stoppages and excessive wear on component contact surfaces during firing; therefore, keep the pistol covered whenever possible.

Corrosion is less likely to form on metal parts in a dry climate. Therefore, lightly lubricate internal working surfaces with CLP. Do not lubricate external parts of the pistol. Wipe excess lubricant from exposed surfaces. Do not lubricate internal components of the magazine.

Heavy Rain and Fording Operations

The following maintenance procedures are followed during periods of heavy rains or during fording operations:

- Always attempt to keep the pistol dry.
- Drain any water from the barrel prior to firing.
- Dry the bore with a swab and cleaning rod.
- Generously lubricate internal and external surfaces of the pistol with CLP.

Amphibious Conditions

If the pistol comes into contact with salt water, clean the pistol as soon as possible. If time does not permit cleaning in accordance with Technical Manual (TM) 1005A-10/1, *Operator's Manual, Pistol Semiautomatic, 9mm, M9*, then wash the pistol with fresh water

Chapter 2
Weapons Handling

Weapons handling is a method of providing consistent and standardized procedures for handling, operating, and employing the M9 service pistol. Understanding and applying the principles of weapons handling are critical to developing safe and consistent weapons skills. Strict adherence to training and diligent practice will make weapons handling instinctive. Mission accomplishment and survival during combat depend on a Marine's ability to react instinctively and with confidence.

Note: The procedures in this chapter are written for right-handed Marines; left-handed Marines should reverse directions as needed.

Safety Rules

The following safety rules are the foundation for responsible weapons handling. They must be observed at all times, both in training and combat.

Safety Rules

Rule 1: Treat every weapon as if it were loaded.
This rule is intended to prevent unintentional injury to personnel or damage to property from handling or transferring possession of a weapon.

Rule 2: Never point a weapon at anything you do not intend to shoot.
This rule enforces the importance of muzzle awareness and reinforces positive identification of the target.

Rule 3: Keep your finger straight and off the trigger until you are ready to fire.
This rule is intended to minimize the risk of firing the weapon negligently (when not firing, the trigger finger is straight along the receiver, outside of the trigger guard). This rule also reinforces positive identification of the target.

Rule 4: Keep weapon on safe until you intend to fire.
This rule enforces the use of the weapon's own safety feature and reinforces positive identification of the target.

Weapons Conditions

The M9 service pistol's level of readiness is defined by three specific conditions. The steps in the loading and unloading process take the pistol through the specific conditions that indicate the pistol's readiness for live fire. The Marine must understand and know the condition of his pistol at all times.

Condition 1. Magazine inserted, round in chamber, slide forward, and safety on.

Condition 2. Not applicable to the M9 service pistol.

Condition 3. Magazine inserted, chamber empty, slide forward, and safety on.

Condition 4. Magazine removed, chamber empty, slide forward, and safety on.

Determining a Weapon's Condition

There are two methods for determining the pistol's condition: checking the round indicator on the right side of the pistol and conducting a chamber check.

Checking the Round Indicator

When there is a round in the chamber, the upper surface of the extractor protrudes from the right side of the slide. The protrusion can be felt by sliding either the thumb or the index finger of the left hand over the top of the slide and across the extractor. See figure 2-1.

Figure 2-1. Checking the Round Indicator.

Conducting a Chamber Check

To conduct a chamber check—

- Point the pistol in a safe direction and grasp the pistol grip with the right hand.
- Place the trigger guard in the palm of the left hand. Use the thumb and index finger to grasp the forward end of the slide at the indentations under the front sight, behind the muzzle. See figure 2-2.

Figure 2-2. Placement of the Left Hand.

Caution

Ensure the muzzle does not cover the hand or fingers.

- Keep thumb in place around the pistol's backstrap and rotate fingers of the right hand over the top of the slide in front of the rear sight.
- Pull the slide to the rear by pushing forward with the right thumb and pulling back on the rear sight with the fingers. Use the left hand to steady the pistol and to assist in pulling the slide to the rear.
- Use the right hand to hold the slide to the rear (just enough to visually inspect the chamber for a round). Physically check for a round by inserting a finger of the right hand into the chamber area. See figure 2-3.

Figure 2-3. Chamber Check.

Note: At night or in low light conditions, the Marine's visibility is reduced; therefore, the Marine will have to rely on the physical check with the finger to determine if a round is in the chamber.

Caution

Pulling the slide too far to the rear while inspecting the chamber may cause a double feed or the ejection of a round.

- Remove the finger from the chamber and release tension on both hands to allow the slide to go forward. Ensure that the slide is all the way forward.

Weapons Commands

Weapons commands direct the Marine to safely load, unload, and employ the M9 service pistol. Six commands are used in weapons handling:

"*Load*"
This command is used to take a weapon from Condition 4 to Condition 3.

"*Make Ready*"
This command is used to take a weapon from Condition 3 to Condition 1.

"*Fire*"
This command is used to engage targets.

"*Cease Fire*"
This command is used to cease target engagement.

"*Unload*"
This command is used to take a weapon from any condition to Condition 4.

"*Unload, Show Clear*"
This command is used to require a second individual to check the weapon to verify that no ammunition is present before the weapon is put into Condition 4.

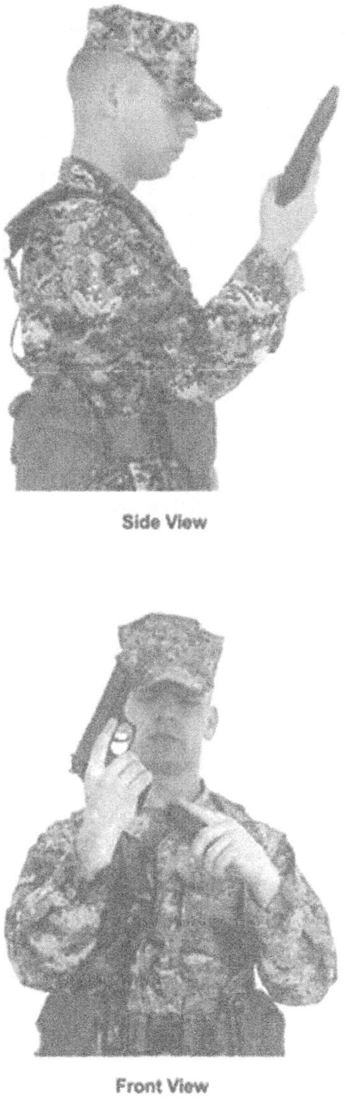

Figure 2-4. Indexing the Magazine.

Loading the Pistol

Perform the following steps to load the pistol (take the pistol from Condition 4 to Condition 3):

- Ensure the pistol is on safe.
- Use the right hand to grip the pistol grip firmly. Ensure that the pistol is pointed in a safe direction, bring the trigger guard to the right of eye level and cant the pistol so the magazine well faces inboard at approximately a 45-degree angle to the deck. Draw the right elbow in to facilitate control of the pistol.
- Use the left hand to remove a filled magazine from the ammunition pocket. Index the magazine by sliding the index finger along the forward edge of the magazine. See figure 2-4.
- Insert the filled magazine into the magazine well by guiding it with the index finger and, with the fingers extended, pushing it in with the heel of the hand until it is fully seated. Do not relinquish contact with the magazine until it is fully seated. See figure 2-5.

Figure 2-5. Seating the Magazine.

Making the Pistol Ready

Perform the following steps to take the pistol from Condition 3 to Condition 1:

- Firmly grip the pistol grip with the right hand. Ensure that the pistol is pointed in a safe direction and the slide is in its forward position.
- Rotate the magazine well outboard to facilitate pulling the slide to the rear. With the fingers and thumb of the left hand, grasp the serrated sides of the slide just forward of the safety. See figure 2-6.

Caution

Ensure the muzzle does not cover the hand or fingers.

- Pull the slide to its rearmost position by pushing forward with the right hand while pulling back on the slide with the left hand.
- Release the slide, this strips a round from the magazine and chambers it as the slide moves forward.
- Ensure the pistol remains on safe.
- Conduct a chamber check to ensure a round is in the chamber.

Note: A chamber check may be conducted at any time to check the pistol's condition.

Fire

Perform the following steps to fire the pistol:

- Keep trigger finger straight and use the right thumb to take the pistol off safe.
- Place the trigger finger on the trigger and apply pressure to the trigger until the shot is fired.

Right Side View

Left Side View

Figure 2-6. Grasping the Slide to Make Ready.

Cease Fire

Perform the following steps to execute a cease fire of the pistol:

- Remove the finger from the trigger and place it straight along the receiver.
- Place the pistol on safe without breaking the grip of the right hand.
- Assume a carry or transport position.

Unloading the Pistol

Perform the following steps to take the pistol from any condition to Condition 4:

- Use the right hand to grip the pistol firmly. Ensure that the pistol is on safe.
- Rotate the pistol so the magazine well is pointed inboard and angled down.

  Note: The angle of the magazine well must allow the magazine to fall freely from the well once the magazine release button is engaged.

- Depress the magazine release button to remove the magazine from the pistol. Catch the magazine with the left hand and retain it.
- Push upward on the slide stop with the right thumb and maintain pressure. Rotate the weapon so the chamber is outboard.

  Note: A left-handed Marine pushes upward on the slide stop with the left index finger.

- Reach over the top of the pistol with the left hand and grasp the slide serrations with the thumb and index finger. The left hand should partially cover the ejection port so it is positioned to catch an ejected round.
- Point the pistol in a safe direction and fully retract the slide and lock it to the rear. At the same time, catch the ejected round with the left hand. See figure 2-7.

Side View

Front View

Figure 2-7. Catching the Ejected Round.

- Rotate the pistol so the inside of the chamber can be seen. Visually inspect the chamber to ensure it is empty.
- Press the slide stop to release the slide and observe it going forward on an empty chamber.

Unloading and Showing the Pistol Clear

Perform the following steps to take the pistol from any condition to Condition 4. See figure 2-8.

Figure 2-8. Unload, Show Clear.

- Use the right hand to grip the pistol firmly. Ensure that the pistol is on safe.
- Rotate the pistol so the magazine well is pointed inboard and angled down.

  Note: The angle of the magazine well must allow the magazine to fall freely from the well once the magazine release button is engaged.

- Depress the magazine release button to remove the magazine from the pistol. Catch the magazine with the left hand and retain it.
- Push upward on the slide stop with the right thumb and maintain pressure. Rotate the weapon so the chamber is outboard.

  Note: A left-handed Marine pushes upward on the slide stop with the left index finger.

- Reach over the top of the pistol with the left hand and grasp the slide serrations with the thumb and index finger. The left hand should partially cover the ejection port so it is positioned to catch an ejected round.
- Point the pistol in a safe direction and fully retract slide and lock it to the rear. At the same time, catch the ejected round with the left hand.
- Rotate the pistol so the inside of the chamber can be seen. Visually inspect the chamber to ensure it is empty.
- Bring the pistol to the administrative transport and have another Marine visually inspect the chamber to ensure that—
  - The chamber is empty, no ammunition is present, and the magazine is removed.
  - The pistol is on safe.
- Acknowledge that the pistol is clear.
- Press the slide stop to release the slide and observe it going forward on an empty chamber.

Emptying the Magazine

Once the pistol is unloaded, the pistol magazine can be emptied of ammunition. To empty the magazine, perform the following steps:

- Hold the magazine upright with the back of the magazine tube against the palm of the hand.
- Push the top round forward with the thumb and catch it with the other hand as it is removed.
- Repeat until the magazine is empty.

Filling the Magazine

Prior to loading the pistol, the pistol magazine must be filled with the prescribed number of rounds of ammunition. See figure 2-9. To fill the magazine, perform the following steps:

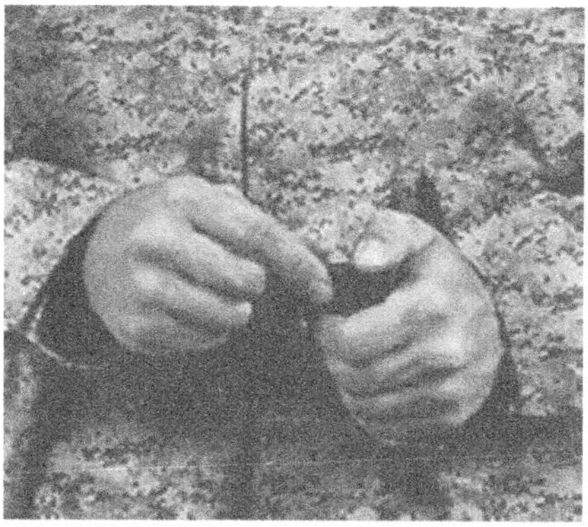

Figure 2-9. Filling the Magazine.

- Hold the magazine with the back of the magazine against the palm of the hand and the follower up.
- Use the other hand to place a round (primer end first) on the follower in front of the magazine lips.
- Press down on the round and slide the round completely back under the lips. The thumb or finger may push down on the back of the round to assist movement. The base of the round should be flush with the back of the magazine.

Repeat this procedure until the magazine is filled with the appropriate number of rounds. Holes on the back of the magazine allow the visual counting of rounds in five-round increments.

Reloading the Pistol

The Marine's ability to reload the pistol quickly improves his chance for success on the battlefield.

Dry Reload

A dry reload is conducted when the pistol runs out of ammunition during engagement and the slide locks to the rear. See figure 2-10. Perform the following steps to conduct a dry reload with the slide locked to the rear:

> *Note:* The pistol is not placed on safe during a dry reload; the trigger finger is taken out of the trigger guard and placed straight along the side of the receiver.

- Seek cover, if the situation permits.

Figure 2-10. Pistol Ran Dry of Ammunition.

- Retain the firing grip with the right hand and pull the pistol in close to the body to facilitate control. Bring the trigger guard to the right of eye level and cant the pistol so the magazine well is facing inboard at approximately a 45-degree angle to the deck. See figure 2-11.

Figure 2-11. Bring the Pistol Toward the Body.

- Press the magazine release button and let the empty magazine fall to the deck. At the same time, unfasten the ammunition pocket to withdraw a filled magazine. See figure 2-12.

  Note: The primary objective during a dry reload is to get the pistol back in action as quickly as possible. Following engagement, retrieve the magazine before moving.

- Grasp the magazine by curling the middle finger and thumb of the left hand around the base of the magazine, with the index finger straight along the ammunition pocket. See figure 2-13.

Figure 2-12. Release Magazine and Unfasten Ammunition Pocket.

Figure 2-13. Grasping the Magazine.

- Index the magazine: as the magazine is being withdrawn from the pocket, the index finger should be along the front of the magazine. See figure 2-14.

Figure 2-14. Indexing the Magazine.

- Rotate the hand up so the magazine is aligned with the magazine well.
- Glance quickly at the magazine well, insert the magazine into the magazine well (see fig. 2-15).

Figure 2-15. Glance at the Magazine Well.

- Bring the eyes back on target at the same time as the heel of the left hand seats the magazine, do not relinquish contact with the magazine. See figure 2-16.

Figure 2-16. Seating the Magazine.

- Roll both hands inward to establish a two-handed grip and press the slide release with the left thumb to allow the slide to move forward, chambering the first round, and present the pistol to the target. See figure 2-17.

Note: A left-handed Marine presses the slide release with his trigger finger.

Figure 2-17. Present Pistol Back to Target.

Figure 2-18. Facilitating Control of Pistol.

Condition 1 Reload

In a Condition 1 reload, a partially-filled magazine is removed from the pistol and replaced with a fully filled magazine. A Condition 1 reload is performed when there is a lull in the action or whenever deemed necessary by the Marine. To perform a Condition 1 reload—

- Retain the firing grip with the right hand and pull the pistol in close to the body to facilitate control. Keep the pistol pointed in the direction of the likely threat. See figure 2-18.
- Withdraw a filled magazine from the ammunition pocket with the left hand. Index the magazine and bring it up to the left of eye level. See figure 2-19.

Figure 2-19. Withdrawing and Indexing a Filled Magazine.

- Slide the index finger to the side of the magazine to grasp the magazine between the index and middle fingers. See figure 2-20.

Figure 2-20. Grasping a Filled Magazine.

- Raise the pistol and bring the trigger guard to the right of eye level and cant the pistol so that the magazine well is facing inboard at approximately a 45-degree angle to the deck.

  Note: The angle of the magazine well must allow the magazine to fall freely from the well once the magazine release button is engaged.

- Press the magazine release button with the right thumb to eject the partially-filled magazine from the magazine well. Grasp the magazine between the index finger and thumb. See figure 2-21.

Figure 2-21. Removing a Partially-Filled Magazine.

- Insert the filled magazine into the magazine well (see fig. 2-22). Use the heel of the hand to ensure it is fully seated (see fig. 2-23).

Figure 2-22. Inserting a Filled Magazine.

Figure 2-23. Seating a Filled Magazine.

- Lower the pistol and point it in the direction of the likely threat.

If time permits, examine the partially-filled magazine to determine the number of rounds remaining. Stow the partially-filled magazine in the ammunition pocket for later use.

Reloading Considerations

If possible, take cover before reloading. Always reload before leaving cover to take advantage of the protection.

When reloading, the first priority is to reload the pistol quickly so that it is ready to fire. During a reload, the Marine focuses on reloading only—not on the enemy.

The next priority is for the Marine to retain the magazine during the reload. However, the combat situation may dictate dropping the magazine to the deck when performing a reload (i.e., dry reload). If time permits (i.e., Condition 1 reload), the Marine picks the magazine up or secures the magazine (e.g., ammunition pocket, flak jacket) before moving to another location.

Remedial Action

The M9 service pistol is an effective and extremely reliable weapon. Proper care and preventive maintenance usually ensures the pistol's serviceability. However, stoppages, while infrequent, do occur. To keep the pistol in action, stoppages must be cleared as quickly as possible through remedial action. A malfunction cannot be corrected through remedial action by the Marine.

Stoppage

A stoppage is an unintentional interruption in the cycle of operation; e.g., the slide not moving forward completely. A stoppage is normally discovered when the pistol will not fire. Most stoppages can be prevented by proper care, cleaning, and lubrication of the pistol.

Many stoppages of the M9 service pistol are caused by shooter error. The Marine must be aware of shooter-induced stoppages in order to avoid them or to quickly identify and correct the stoppage and return the pistol to action. In a shooter-induced stoppage, the Marine—

- Fails to make ready.
- Fails to take the pistol off safe prior to firing.
- Engages safety while firing.
- Engages magazine release button while firing.
- Engages slide stop while firing (particularly prevalent when firing with an Isosceles grip).
- Fails to reset the trigger.
- Fails to recognize the pistol has run dry and the slide has locked to the rear.

Malfunction

A malfunction is a failure of the pistol to fire satisfactorily or to perform as designed (e.g., a broken front sight that does not affect the functioning of the pistol). A malfunction does not necessarily cause an interruption in the cycle of operation. When a malfunction occurs, the pistol must be repaired by an armorer.

Remedial Action

There is no one set of procedures (i.e., immediate action) that can be performed to clear all or even most of the stoppages that can occur with the M9 service pistol. Therefore, remedial action requires investigating the cause of the stoppage, clearing the stoppage, and returning the pistol to operation. When performing remedial action, the Marine should seek cover if the tactical situation permits. Once a pistol ceases to fire, the Marine must visually or physically observe the pistol to identify the problem before it can be cleared:

> *Note:* The steps taken to clear the pistol are based on what is observed.

- Remove the finger from the trigger and place it straight along the receiver.
- Bring the pistol in close to the body and in a position to observe the chamber.
- Pull the slide to the rear while observing the chamber area to identify the stoppage. See figure 2-24 on page 2-14.

> *Note:* Ensure the pistol does not move to safe when pulling the slide to the rear.

Figure 2-24. Observing Chamber.

- Correct the stoppage:
  - If there is a round in the magazine but not in the chamber (see fig. 2-25), the slide is released and a round is observed being chambered.

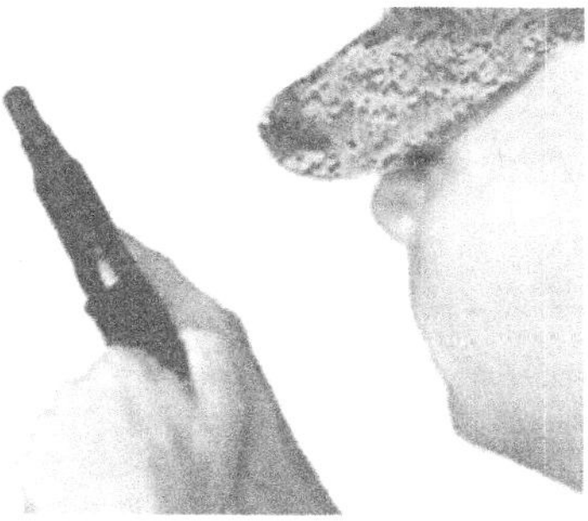

Figure 2-25. Round in Magazine but Not in Chamber.

  - If a round being chambered is not observed, the bottom of the magazine is tapped to seat it properly, and the slide is racked to the rear. See figure 2-26.
  - If there is no round in the magazine or chamber, a reload is conducted. See figure 2-27.

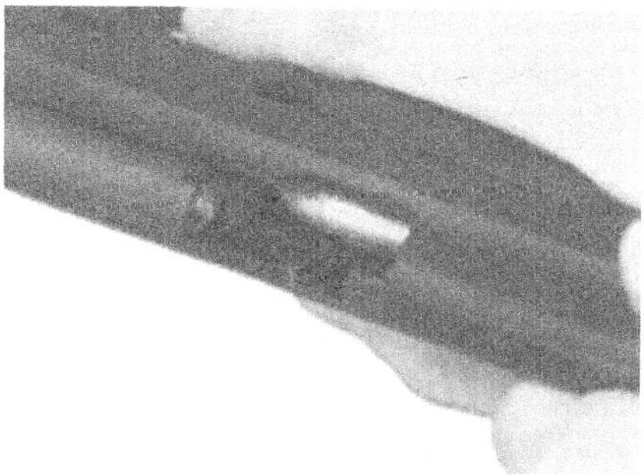

Figure 2-26. Round Not Being Chambered.

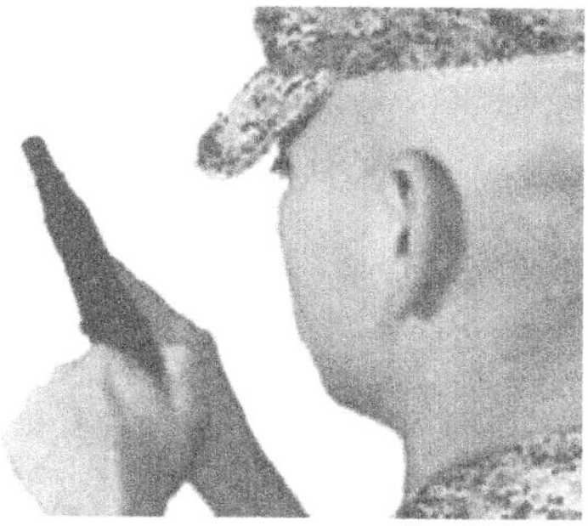

Figure 2-27. No Round in Magazine or Chamber.

- Fire the pistol.

Audible Pop or Reduced Recoil

WARNING

When an audible pop or reduced recoil is experienced, the Marine DOES NOT perform remedial action unless he is in a combat environment. An audible pop occurs when only a portion of the propellant is ignited. It is normally identifiable by reduced recoil and the pistol will not cycle. Sometimes, it is accompanied by excessive smoke escaping from the chamber area.

Training Environment

If an audible pop or reduced recoil is experienced during firing, cease fire immediately. Do not apply remedial action; instead, perform the following steps:

- Remove the finger from the trigger and place it straight along the receiver.
- Point the pistol down range.
- Place the pistol on safe.
- Raise a hand to receive assistance from available range personnel.

Combat Environment

The tactical situation may dictate correction of an audible pop or reduced recoil. To clear the pistol, perform the following steps:

- Remove the finger from the trigger and place it straight along the receiver.
- Seek cover if the tactical situation permits.
- Unload the pistol, but leave the slide locked to the rear.
- Insert something into the bore and clear the obstruction.
- Observe the barrel for cracks or bulges.
- Reload the pistol.

Weapons Carries

As the threat level increases, so should the Marine's readiness for engagement. Weapons carries are designed to place the Marine in a state of increased readiness as the threat level increases. There are two carries with the pistol: the Alert and the Ready. The carries permit quick engagement when necessary.

Alert

The Alert is used when enemy contact is likely (probable). See figure 2-28. The Marine performs the following steps to assume the Alert:

Figure 2-28. Alert.

- Ensure the pistol is on safe.
- Grasp the pistol grip firmly with two hands. The trigger finger is straight and the right thumb is on the safety and in a position to

- Extend the arms down at approximately a 45-degree angle to the body or bend the elbows. See figure 2-29.
- The muzzle of the pistol is pointed in the likely direction of the threat.

Ready

The Ready is used when there is no target, but contact with the enemy is imminent. The Marine performs the following steps to assume the Ready (see fig. 2-30):

- Ensure the pistol is on safe.
- Grasp the pistol firmly with two hands. The trigger finger is straight and the right thumb is on the safety and in a position to operate it.
- Extend arms and raise the pistol to just below eye level so a clear field of view is maintained.
- Point the muzzle of the pistol in the direction of enemy contact.

Weapons Transports

The M9 service pistol is transported in either the holster transport or the administrative transport.

Holster Transport

The holster transport is the most common method of carrying the pistol because it can be transported safely in the holster. This transport is used when there is no immediate threat (enemy contact is remote). See figure 2-31. To transport the pistol in the holster:

- Point the pistol in a safe direction.

WARNING

Ensure the pistol is pointed in a safe direction at all times and does not cover any part of the body while holstering.

Figure 2-29. Alert—Close Quarters.

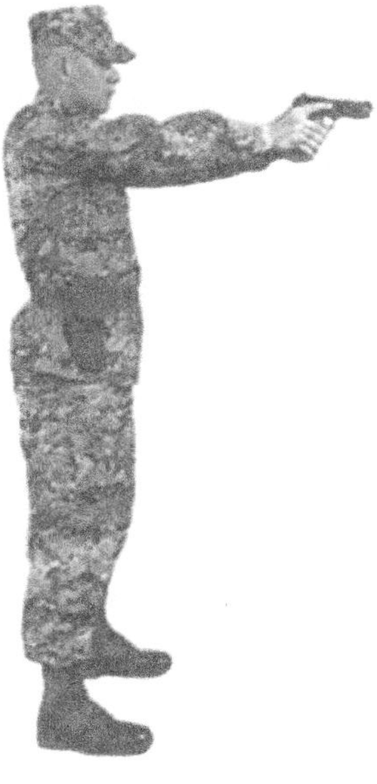

Figure 2-30. Ready.

Figure 2-31. Holster Transport.

Figure 2-32. Administrative Transport.

- Ensure that the safety is on, the slide is forward, and the trigger finger is straight.
- Use the right hand to firmly grip the pistol grip and place the pistol in the holster:
  - Lift the flap of the holster with the left hand.
  - Look down at the holster, bring the pistol back to a position above the holster, and rotate the muzzle down into the holster.
- Push the pistol snugly into the holster and fasten the flap with the right hand.

Administrative Transport

The administrative transport is used to transport the pistol when the Marine does not have a holster. See figure 2-32. The Marine performs the following steps to assume the administrative transport:

- Establish a firm grip around the pistol grip with the right hand.
- Ensure the pistol is on safe, the magazine is removed, the slide is locked to the rear, and the trigger finger is straight along the receiver.
- Bend the elbow to approximately a 45-degree angle so the pistol is positioned near shoulder level. The wrist should be straight so the pistol's muzzle points up.

Combat Mindset

In a combat environment, the Marine must be constantly prepared to engage targets. When a target presents itself, there may be little time to react. The target must be engaged quickly and accurately. It is not enough to simply know marksmanship techniques, the Marine must be able to react instinctively.

The development of a combat mindset can be associated with the carries and holster transport for the pistol. The use of each carry/transport is dictated by the perceived level of threat. Each carry and transport should signify a stage of mental and physical preparedness for combat. The intensity of the Marine's mental and physical preparation depends on the likelihood of enemy contact.

Physical and Mental Preparation

Physical Preparation

In combat, targets can present themselves without warning. Therefore, it is essential for the Marine to maintain proper balance and control of the pistol at all times so that the pistol can be presented quickly and the target engaged accurately.

Speed alone does not equate to effective target engagement. The Marine should fire only as fast as he can fire accurately, never exceeding his physical ability to apply the fundamentals of marksmanship. To be effective in combat, the Marine must train to perfect the physical skills of shooting so they become second nature. The more physical skills that can be performed automatically, the more concentration that can be given to the mental side of target engagement.

Mental Preparation

While combat is unpredictable and constantly changing, the Marine can prepare mentally for the contingencies of the operational setting and confrontation with a threat. The stress of combat, coupled with the limited time available to engage targets, requires concentration on the mental aspects of target engagement; e.g., identification of targets, shoot/no-shoot decisionmaking, and the selection and use of cover. Minimizing stress and maximizing the limited time available to engage targets can be accomplished by—

- Knowing the combat environment and being constantly aware of the surroundings (e.g., terrain, available cover, possible threats) enables the Marine to quickly present the pistol and accurately engage targets.
- Identifying and evaluating possible courses of action and developing potential plans for target engagement that will be appropriate to the combat situation.
- Instilling confidence in a Marine's ability to fire well-aimed shots in the stress of a combat situation. A key factor in a Marine's level of confidence is the degree to which he has mastered the tactics, techniques, and procedures of pistol marksmanship.

Threat Levels

No Immediate Threat

When there is no immediate threat, the Marine assumes the holster transport. The pistol should be in Condition 1. This is the lowest level of awareness for the Marine in a combat environment, but the Marine must stay alert and aware of any nearby activity. To prepare for target engagement at this level, the Marine must—

- Be aware of likely areas of enemy contact.
- Be aware of the condition of his pistol.
- Establish a plan or course of action to present the pistol to a target should a target appear.
- Mentally review appropriate actions such as reloading and remedial action.

Contact Likely (Probable)

If enemy contact is likely (probable), the Marine assumes the Alert. When enemy contact is likely, the Marine should—

- Expect enemy contact and be constantly prepared to present the pistol.
- Search the entire area for indications of enemy targets and for suitable terrain features that offer cover and concealment. The Marine should avoid restricting the search to a single terrain feature because this hinders awareness to a sector of the battlefield and to approach by the enemy.
- Be mentally prepared for contact. Plan a course of action for immediate response to a target. Modify the plan of action as needed.

- Be physically prepared to fire. Maintain proper balance at all times. Avoid self-induced physical fatigue. For example, do not grip the pistol so tightly that fingers, hands, and arms tire from carrying the pistol.

Contact Imminent

When contact with an enemy target is imminent, assume the Ready. In this carry, the Marine is at the highest level of awareness and is constantly searching for and expecting a target. To fire well-aimed shots upon target detection, the Marine must be at the peak of his mental preparation, all distractions must be eliminated, and his focus must be on firing an accurate shot. In the Ready, the Marine must—

- Keep the pistol oriented in the general direction of observation (eyes, muzzle, target).
- Maintain a clear field of view above the pistol sights until the target is detected.
- Be mentally and physically prepared to engage the target. The Marine must be ready to:
  - Identify the target.
  - Sweep the safety.
  - Apply the fundamentals of marksmanship.
- Move only as fast as he is capable of delivering well-aimed shots, ensuring that speed of engagement does not exceed his physical abilities.
- Search the entire area for indications of enemy targets, lowering the pistol enough to observe a clear field of view of the area.

Transferring the Pistol

The Marine's ability to transfer a pistol to another Marine is critical to safe weapons handling. There are two methods for transferring the pistol from one Marine to another: show clear transfer and condition unknown transfer. Each transfer is performed based on the operational environment/combat situation.

Show Clear Transfer

To transfer the pistol—

- Grasp the pistol firmly in the right hand while ensuring that the pistol is on safe.
- Remove and retain the magazine.
- Lock the slide to the rear and catch the round if there is a round in the chamber.
- Inspect the chamber visually to ensure it is empty and leave the slide locked to the rear. To transfer the pistol—
  - If the receiving Marine is to the right: Cradle the trigger guard in the palm of the left hand and wrap the fingers around the top of the pistol. Release the firing grip.
  - If the receiving Marine is to the left: With the left hand, grasp the slide of the pistol with the thumb over the slide and the fingers underneath. Release the firing grip.
- Ensure that the muzzle points up at a 45-degree angle in a safe direction and the chamber exposed. Hand the pistol to the other Marine, grip first. See figure 2-33.

Figure 2-33. Show Clear Transfer.

- The receiving Marine—
  - Grasps the pistol grip with the trigger finger straight along the receiver.
  - Inspects the chamber visually to ensure it is empty.
  - Ensures the pistol is on safe.

Condition Unknown Transfer

To transfer the pistol—

- Grasp the pistol firmly in the right hand while ensuring that the pistol is on safe. To transfer the pistol—
  - If the receiving Marine is to the right: Cradle the trigger guard in the palm of the left hand and wrap the fingers around the top of the pistol. Release the firing grip. See figure 2-34.

Figure 2-34. Condition Unknown Transfer to the Right.

  - If the receiving Marine is to the left: With the left hand, grasp the slide of the pistol with the thumb over the slide and the fingers underneath. Release the firing grip. See figure 2-35.

Figure 2-35. Condition Unknown Transfer to the Left.

- Ensure that the muzzle points up at a 45-degree angle in a safe direction. Hand the pistol to the receiving Marine, grip first.
- The receiving Marine—
  - Grasps the pistol grip with the trigger finger straight along the receiver.
  - Ensures the pistol is on safe.
  - Conducts a chamber check to determine the condition of the pistol.
  - Remove the magazine and count the number of rounds in the magazine by using the counting holes, if time permits. Re-insert the magazine into the magazine well ensuring it is fully seated.

CHAPTER 3
FUNDAMENTALS OF PISTOL MARKSMANSHIP

The fundamentals of pistol marksmanship are aiming, trigger control, and breath control. Understanding and applying the basic pistol marksmanship fundamentals ensures the Marine's effectiveness in target engagement. The fundamentals must be continually studied and practiced because they are the means by which accurate shots are placed on target. A Marine with a solid foundation in the fundamentals of marksmanship will be successful in the application of these fundamentals during combat.

Aiming

Maintaining the correct relationship between the pistol sights is essential for accurate target engagement. Because of the short distance between the pistol sights, a small error in their alignment causes a considerable error at the target.

Sight Alignment

Sight alignment is the relationship between the front sight and rear sight with respect to the aiming eye. Correct sight alignment is the front sight centered in the rear sight notch with the top edge of the front sight level **aligned** with the top edge of the rear sight. There should be equal space on either side of the front sight. See figure 3-1.

Figure 3-1. Sight Alignment.

Establishing Sight Alignment

The pistol is fired without benefit of bone support; therefore, the pistol is in constant motion. The Marine must understand this, yet continually strive to align the sights. To fire accurately, the sights must be aligned when the shot breaks.

Grip

The grip is key to acquiring sight alignment. If the grip is correct, the front and rear sights should align naturally. Dry fire during presentation of the M9 service pistol aids in obtaining a grip that allows sight alignment to be acquired consistently.

Controlled Muscular Tension

There must be enough controlled muscular tension in the grip, wrists, and forearms to hold the pistol steady and level the barrel to maintain sight alignment. Consistent tension stabilizes the sights and maintains sight alignment.

Sight Picture

Sight picture is the placement of the front sight in relation to the target while maintaining sight alignment. See figure 3-2 on page 3-2.

Because the pistol is constantly moving, sight picture is acquired within an aiming area that is located center mass on the target. The aiming area allows for movement of the sights on the target while maintaining sight alignment. Each Marine defines an acceptable aiming area within his own ability to stabilize the sights. Time, distance to the target, and personal ability affect dictate the aiming area. As the Marine becomes more proficient with the pistol, the aiming area becomes more precise.

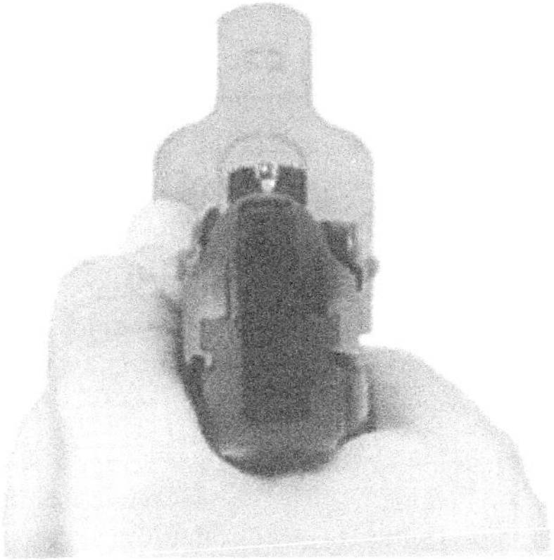

Figure 3-2. Sight Picture.

The aiming area is determined by the Marine's stability of hold. The proper grip stabilizes the sights so sight alignment can be maintained, but the sights move continuously within the aiming area of the target. The Marine understands the pistol's movement and learns to apply trigger control as he is obtaining sight alignment/sight picture within the aiming area so the shot breaks the moment sight picture is established.

Relationship Between the Eye and the Sights

The human eye can focus clearly on only one object at a time. The Marine must focus on the top edge of the front sight and fire the shot while maintaining the relationship between the front and rear sights within the aiming area. Focusing on the top edge of the front sight rather than the target keeps the front sight clear and distinct, which allows the Marine to detect minor variations in sight alignment. Secondary vision allows the Marine to see the target (although slightly blurred) and maintain sight picture within his aiming area.

Trigger Control

Trigger control is the Marine's skillful manipulation of the trigger that causes the pistol to fire while maintaining sight alignment and sight picture. Proper trigger control aids in maintaining sight alignment while the shot is fired.

Sight Alignment and Trigger Control

Aiming and trigger control are mutually supportive—one cannot be performed without the other. Sight alignment and trigger control must be performed simultaneously to fire an accurate shot. As pressure is applied to the trigger, the sights may move, causing them to be misaligned. To fire accurate shots, the sights must be aligned when the shot breaks. Trigger control can actually assist in aligning the sights. With proper trigger finger placement and consistent muscular tension applied to the grip, the sights can be controlled as the trigger is moved to the rear. If the sights move extensively while pressing the trigger, this can indicate an improper grip or inconsistency in the muscular tension being applied to the grip.

Grip

A firm grip is essential for good trigger control. The grip is established before applying trigger control and is maintained throughout the firing process. To establish the grip, the hand is placed around the pistol grip in a location that allows the trigger finger to move the trigger straight to the rear while maintaining sight alignment. Once the grip is established, it should be firm enough to allow manipulation of the trigger while maintaining sight alignment. The pressure applied to the grip must be equal to or more than the pressure required to move the trigger to the rear. If the pressure is not applied correctly, the sights move as the trigger is pressed to the rear and sight alignment is disturbed.

Trigger Finger Placement

Once the grip is established, the finger is placed on the trigger. Placement of the finger should be natural and allow free movement of the trigger finger. A natural trigger finger placement allows the trigger to be moved straight to the rear while maintaining sight alignment. If the finger presses the trigger to the side, it can cause an error in sight alignment and shot placement.

Each Marine must experiment with finger placement in order to determine effective placement on the trigger. Once established, effective trigger finger placement allows the trigger to be consistently moved straight to the rear while maintaining sight alignment.

Types of Trigger Control

Uninterrupted Trigger Control

During uninterrupted trigger control, the Marine applies a steady, unchanging pressure to the trigger until the shot is fired. Uninterrupted trigger control is particularly effective at close range, when the target area is large, and when stability of hold is not critical for accuracy. To apply uninterrupted trigger control, apply pressure on the trigger while maintaining focus on the top edge of the front sight. Continue pressure on the trigger to begin moving the trigger straight to the rear while obtaining sight alignment and sight picture. Move the trigger straight to the rear in a single, smooth motion with no hesitation.

Interrupted Trigger Control

Interrupted trigger control is particularly effective at longer ranges, when the target is small, and when stability of hold is critical to maintaining sight picture in the aiming area. This method is also used if the pistol sights move outside the aiming area when applying trigger control. If the sight picture is outside the aiming area, the Marine stops and holds the rearward movement on the trigger until sight picture is re-established. When sight picture is re-established, the rearward movement of the trigger is continued until the shot is fired.

Breath Control

Breathing causes movement of the chest, abdomen, and shoulders, which causes the pistol sights to move vertically while attempting to aim and fire. Therefore, it is necessary to stop breathing for a period of time while firing a shot or a series of shots.

The object of breath control is to stop breathing just long enough to fire the shot while maintaining sight alignment, stabilizing the sights, and establishing the sight picture. To be consistent, the breath should be held at the same point in the breathing cycle; i.e., the natural respiratory pause.

Breathing should not be stopped for too long because it has adverse visual and physical effects. Holding the breath longer than is comfortable results in a lack of oxygen that causes vision to deteriorate and then affects the ability to focus on the sights.

Application of Marksmanship Fundamentals in Field Firing

Compression of the Fundamentals

Pistol engagements typically occur over close distances and are short in duration. Because an immediate response to the threat is required, the application of the fundamentals must be a conditioned response that is executable in a compressed time. The goal of successful, quick target engagement is the application of the fundamentals of marksmanship while firing the shot the moment weapon presentation is complete.

The time required to move the trigger to the rear while acquiring and maintaining sight alignment and sight picture is unique to each Marine and is based on his capabilities. Each Marine should know his abilities and fire only as quickly as he is capable of firing accurately. The Marine must not exceed his shooting skills in an effort to get rounds off quickly.

Aiming

In field firing, the fundamentals are applied in a compressed time so sight alignment and sight picture are achieved as the shot is fired. Although the target must be quickly engaged in combat, sight alignment is still the first priority: strive for a clear front sight. Distance to the target and the size of the target affects sight alignment as follows:

- As the distance to the target increases and the size of the target decreases, sight alignment becomes more critical to target engagement. Accurate sight picture/sight alignment cannot be compromised for speed.
- Sight alignment is critical to the effective engagement of smaller targets such as partially exposed targets.
- As the distance to the target decreases, perfect sight alignment is not as critical, but there must be a relationship between the sights and their placement on the target within the aiming area to ensure accuracy.

Trigger Control

Proper trigger control aids in maintaining sight alignment while the shot is fired. As pressure is applied to the trigger, the sights may move, causing them to be misaligned. Therefore, the Marine is constantly re-aligning the sights as pressure is applied to the trigger. Sight alignment and trigger control must be performed simultaneously to fire an accurate shot.

Breath Control

During combat, the Marine's breathing and heart rate often increase due to physical exertion or the stress of battle. The key to breath control in field firing is to stop breathing just long enough to fire an accurate shot or a series of shots.

CHAPTER 4
PISTOL FIRING POSITIONS AND GRIP

The M9 service pistol is fired from the standing, kneeling, and prone positions. Each firing position may be adapted to either a Weaver or Isosceles variation, each possessing a distinct advantage in combat. The Weaver variation stabilizes the pistol sights. The Isosceles variation manages recoil. The advantages apply whether the Marine is firing in the standing, kneeling, or prone position. The Marine must select and assume a stable firing position that provides a solid foundation for accurate shooting while meeting the demands of the combat situation

Note: The procedures in this chapter are written for right-handed Marines; left-handed Marines should reverse directions as needed.

The Pistol Firing Position

During combat, the Marine selects a firing position based on mobility, observation of the enemy, and stability.

Mobility

A firing position must provide mobility should the Marine need to move. The standing position permits maximum mobility because it is quickly assumed and easily maneuvered from and it permits lateral mobility to engage widely dispersed targets. The prone position provides limited mobility because it is the most time-consuming position to get into and out of and it lacks the lateral mobility required to engage dispersed targets.

Observation of the Enemy

A firing position should allow observation of the enemy while minimizing the Marine's exposure. During combat there can be many obstructions to a clear field of view. Terrain features such as vegetation, earth contours, and manmade structures often dictate the firing position. The prone position normally allows the least exposure, but it usually provides a limited field of view. Kneeling may provide a wider field of view, but generally provides less concealment.

Stability

A solid firing position establishes a stable foundation for target engagement and provides accurate and consistent shooting. The definition of a stable position is one in which the body is positioned so as to resist forces that cause motion (i.e., recoil and movement of the pistol sights). The standing position is the least stable firing position, while the prone is the most stable firing position. A consistent, stable position is assumed for two distinct purposes:

- Minimize the pistol's movement in order to control the sights and to deliver accurate fire on a target.
- Minimize the affects of recoil in order to recover the sights to the same area on the target.

During combat, it may be necessary to engage the same target more than once to eliminate it. If the firing position is stable, the pistol's sights should recover to the same area on the target, allowing rapid re-enactment. Distributing the body's weight to balance the position also stabilizes the pistol and allows better management of recoil. A pistol firing position is stabilized through controlled muscular tension. Because the pistol is fired without benefit of bone support, muscular tension is needed in the body to stabilize the position and the pistol sights. Muscular tension must follow these guidelines—

- A consistent amount of muscular tension is needed to hold the pistol steady so the sights are aligned with the aiming eye and the target.

- Controlled and consistent tension in the body provides resistance that the Marine uses to manage recoil and bring the sights back on target quicker. However, too much tension can cause strain or trembling.
- Muscular tension is correct when the Marine can control the pistol before, during, and after firing the shot.

Pistol Firing Grip

A proper grip is one that provides maximum control of the pistol before, during, and after firing. It stabilizes the pistol sights before firing, allows trigger control to be applied during firing, and manages recoil after firing.

Before Firing: Stabilize the Sights

To fire an accurate shot, the sights must be stabilized prior to and as the bullet exits the muzzle of the pistol. A proper grip controls the alignment of the pistol's sights and stabilizes the sights so an accurate shot may be fired. The following guidelines apply to the establishment of a proper grip—

- There must be muscular tension in the wrist and forearms. Consistent muscular tension in the wrist, forearms, and grip helps maintain sight alignment by reducing the movement in the grip that can cause movement in the sights. The grip is correct when it allows the sights to be naturally aligned to the aiming eye.
- When establishing a two-handed grip, equal pressure must be applied with both hands. Consistent, equal pressure from both hands stabilizes the sights and allows them to be aligned and level with respect to the aiming eye.
- The hand grips as high on the back strap as possible, bringing the centerline of the bore as close as possible to a parallel line with the bones in the forearm.

During Firing: Allow Trigger Control

The Marine's grip provides a foundation for the movement of the trigger finger. The trigger finger applies positive pressure on the trigger as an independent action, completely free of the other muscles of the gripping hand. The Marine should not apply excessive pressure to the web of the hand touching the pistol's back strap because excessive pressure interferes with the manipulation of the trigger by the trigger finger.

After Firing: Manage Recoil

Once a shot is fired, the pistol recoils, disturbing alignment of the sights. A proper grip facilitates a quick recovery from recoil so the sights quickly return to the same area on the target. The Marine's grip determines the following during recoil:

- The amount the muzzle climbs during recoil depends on the amount of controlled muscular tension in the grip and wrists applied to stabilize the pistol and create consistency in resistance to recoil. Controlled muscular tension allows the sights to recover consistently back on target within a minimum amount of time.
- Firm, equal pressure must be applied to the grip with both hands to ensure that the pistol does not slip during recoil. An improper grip or lack of controlled muscular tension causes the pistol to move after the shot is fired, disrupting sight alignment and requiring the Marine to re-establish his grip.
- Locking out the wrist of the firing hand, similar to executing a punch, helps provide resistance to recoil and speed recovery. The elbows should be at a position slightly less than locked out to help absorb the recoil and aid in recoil management.

Withdrawing the Pistol From the Holster

The firing grip is not established in the holster, however, a proper firing grip can be assumed quickly if the pistol is withdrawn from the holster correctly. The pistol should be withdrawn from the holster in one continuous, fluid motion:

- Place the heel of the left hand at the center of the torso with the fingers extended toward the target. At the same time, unfasten and release the D-ring with the right hand. See figure 4-1.

Figure 4-1. Withdrawing the Pistol From the Holster—Step 1.

Note: The left hand is placed on the torso in a position that allows a two-handed firing grip to be established in a minimum amount of movement.

- Place the right thumb on the forward edge of the lower portion of the holster and the fingers around the back edge of the holster, keeping the trigger finger straight. See figure 4-2.

Note: The left hand may assist in holding the holster flap up.

Figure 4-2. Withdrawing the Pistol From the Holster—Step 2.

- Slide the hand up the holster until the fingers come in contact with the pistol grip. At the same time, keep the thumb above the pistol to guide the holster flap up. See figure 4-3 on page 4-4.
- Grasp the pistol grip with the fingers and draw the pistol straight up. Continue withdrawing the pistol while moving the thumb to a position on the safety. See figure 4-4 on page 4-4.

Note: This hand placement allows a firing grip to be established once the thumb disengages the safety. Any adjustments made to the firing grip after the safety is disengaged should be minor.

Figure 4-3. Withdrawing the Pistol From the Holster—Step 3.

Figure 4-4. Withdrawing the Pistol From the Holster—Step 4.

- Once the muzzle clears the holster, rotate the muzzle forward to clear the body. Ensure the muzzle is pointed in a safe direction.
- Establish a two-handed grip on the pistol by joining the left hand with the right hand in the front of the torso. See figure 4-5.

Caution

Ensure the muzzle does not cover the left hand when establishing the two-handed grip.

Weaver and Isosceles Variations

The size of the target, distance to the target, time, and type of engagement needed (i.e., two shots, single precision shot) determine whether to fire in the Weaver or Isosceles variation. The Weaver and Isosceles variations each consist of three firing positions: standing, kneeling, and prone. The firing position chosen is based on the combat situation and the Marine's body configuration, and

Figure 4-5. Withdrawing the Pistol From the Holster—Step 5.

it must permit balance, control, and stability during firing. In addition, there is a specific firing grip that supports each position and the combination of the grip and the body configuration is key to establishing the variation. See table 4-1 for a list of advantages and disadvantages of both the Weaver and the Isosceles variations.

At longer ranges, the target is smaller, a more precise shot is required to eliminate the target, and any small movement moves the sights off the target; therefore, the pistol must be steadied. Although the Weaver variation is effective at any distance, some aspects of the position make it more effective for long range or precision shots on small or partially exposed targets; i.e., sight alignment and sight picture are easier for stability of hold.

The Isosceles variation is effective at any distance, however, some aspects of the position make it more effective for close range engagements. When confronted with a target, the natural physical reaction is to face the target and push out with the arms. This makes the Isosceles variation advantageous for quick engagements at close range. When a target is at close range, it must be engaged quickly before it engages the Marine. The management of recoil is a bigger factor in close-range engagements because it is more likely that multiple shots will be fired to eliminate the target and the sights have to recover quickly back on target. However, at close ranges, the target is larger so stability of hold is not as important because it is easier to hold the sights on target and sight picture is not as critical.

Standing Position

The standing position is the most often employed position during a pistol engagement due to the short distance of the engagement and the nature of combat. When properly assumed, the standing position provides a stable base for firing, a clear field of view, and excellent mobility. The standing position can be adapted to either the Weaver variation or the Isosceles variation.

Weaver Standing Position

The key to successful employment of the Weaver variation is the body's angle to the target and the push-pull pressure applied to the grip. See figure 4-6.

Table 4-1. Advantages and Disadvantages.

Variation	Advantages	Disadvantages
Weaver	Additional balance, control, and stability of hold during firing due to placement of the arms (left arm bent, pistol is in close to the body). Easier to maintain sight picture because the pistol's foundation is steady.	Recoil has a greater impact due to the hand placement on the pistol (some of the pistol grip is exposed and pressure is applied in two different directions around the pistol). Recovery of the sights back on target may take longer since recoil is affected.
Isosceles	Recoil has a lesser impact because muscular tension and grip pressure are evenly distributed around the pistol. Allows quicker recovery of the sights on target.	Stability of hold is degraded due to the pistol being further from the body without support. Since it is harder to steady the pistol, acquiring sight picture becomes more difficult.

To assume the Weaver standing position—

- Face the target and make a half turn to the right, keeping the pistol oriented toward the target. This orients the body at approximately a 40 to 60 degree angle oblique to the target. The shoulders are angled to the target, the left shoulder forward of the right. The feet are about shoulder-width apart, the left foot forward of the right.
- Grip the pistol grip firmly with the right hand. Place the right thumb on the safety.
- Keep the shoulders at a 40 to 60 degree angle oblique to the target and raise the right arm and extend it across the body toward the target. Ensure that the right shoulder does not roll forward or turn toward the target.
- Extend the left arm to the target, bending the left elbow to join the left and right hands.

Figure 4-6. Weaver Standing Position.

The left elbow is inverted and tucked in toward the body so the left arm supports the pistol.

Note: The angle of the body determines how much the elbow bends.

- Establish a two-handed firing grip in the Weaver variation. See figure 4-7.
- Place the palm of the left hand over the front of the right hand so the palm covers the curled fingers of the right hand. The trigger guard should rest in the "V" formed by the left thumb and forefinger. The knuckles of the left hand are just outboard of the trigger guard. A portion of the pistol grip is exposed.
- Rest the trigger finger naturally, straight and outside of the trigger guard, so the finger can be moved quickly and easily to the trigger.
- Rest the left thumb against the receiver so that both thumbs are on the left side of the pistol. Once the safety is disengaged with the right thumb, the left thumb is placed over the right thumb and positive pressure is applied to hold the right thumb in place.
- Apply rearward pressure with the left hand and forward pressure with the right hand to achieve a "push-pull" grip. Isometric tension (push-pull) stabilizes the pistol during firing.

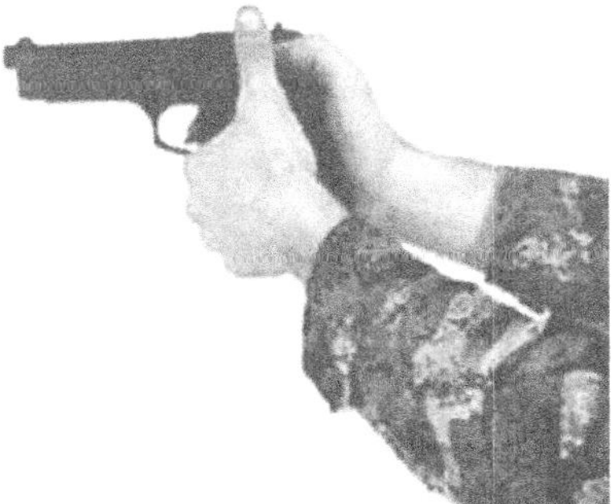

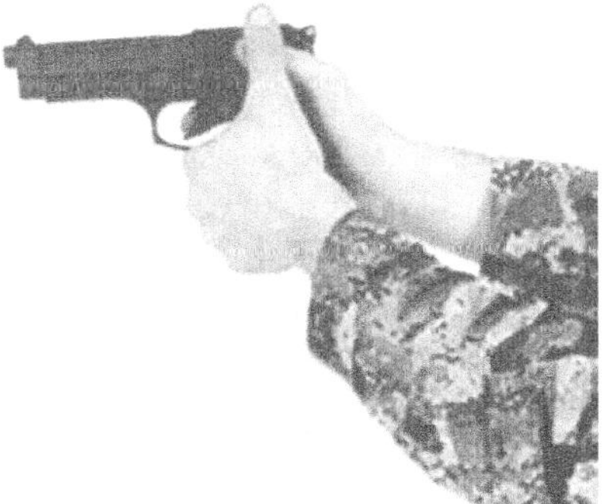

Figure 4-7. Weaver Grip.

- Lean forward slightly and apply muscular tension throughout the body to stabilize the position and manage recoil. The muscular tension in the upper body is not symmetrical due to the "push-pull" tension applied on the grip.
- Keep the head erect so the aiming eye can look through the sights.

Isosceles Standing Position

The key to the Isosceles variation is that the body is squared to the target and equal pressure is applied on the pistol from the grip. To assume the Isosceles standing position—

- Face the target with feet approximately shoulder width apart. The shoulders are squared to the target.
- Establish a two-handed firing grip in the Isosceles variation. See figure 4-8.

Figure 4-8. Isosceles Grip.

- Grip the pistol grip firmly with the right hand. Place the right thumb on the safety.
- Place the heel of the left hand on the exposed portion of the pistol grip in the pocket formed by the fingertips and heel of the right hand. There should be maximum contact between the pistol grip and the hands. Wrap the fingers of the left hand over the fingers of the right hand. Ensure both thumbs rest on the left side of the pistol and point toward the target.
- Apply equal pressure on both sides of the pistol to allow for the best management of recoil.

  Note: Ensure the left thumb does not apply excessive pressure to the slide stop or slide.

- Rest the trigger finger naturally, straight and outside of the trigger guard, so the finger can be moved quickly and easily to the trigger.

  Notes: Index finger of the left hand may or may not rest on the front of the trigger guard.

  Marines with large hands must ensure that their right thumb does not rest on the slide stop, preventing reliable pistol operation.

- Elevate and extend the arms toward the target.
- Roll the shoulders forward and shift the body weight slightly forward to stabilize the position and better manage recoil. The left foot may be slightly forward of the right foot to balance the position. There should be an equal amount of muscular tension on both sides of the body to best manage recoil.
- Tuck the head between the shoulders; the head is extended forward but kept erect so the aiming eye can see through the sights. See figure 4-9.

Figure 4-9. Isosceles Standing Position.

Kneeling Position

The kneeling position offers a smaller exposure than the standing position and greater stability. Increased stability makes the kneeling position effective for longer range shooting. It does not, however, offer as much mobility for quick reaction as the standing position. The kneeling position can be quickly assumed and allows firing from various types of cover. Depending on the cover and the need for observation, the kneeling position may be adapted to a high kneeling, a medium kneeling, a low kneeling, or a two-knee kneeling position. The kneeling positions can be adapted to either the Weaver variation or the Isosceles variation.

Weaver Kneeling Position

The advantage of the Weaver variation of the kneeling position is that it provides bone support due to the left elbow's placement on the knee. The Weaver variation further enables firing from the side of cover while exposing less of the body to a threat. To assume the Weaver kneeling position, the following steps are basic to all adaptations:

- Make a half turn to the right, drop the right foot back or step forward with the left foot, and place the right knee on the deck. The body is positioned at a 40 to 60 degree angle oblique to the target.
- Blade the shoulders at a 40 to 60 degree oblique angle to the target, the left shoulder forward of the right.
- Extend the arms toward the target.
- Bend forward at the waist to better manage recoil.
- Place the flat part of the upper left arm, just above the elbow, in firm contact with the flat surface formed on top of the bent knee. The point of the left elbow extends just slightly past the left knee. However, depending on the need for stability or observation of the enemy, the elbow does not have to rest on the knee.

The following steps provide specifics for each adaptation:

- High kneeling: the toes of the right foot are curled and in contact with the deck or the inside of the foot may be in contact with the deck. Depending on the need for observation, the buttocks may or may not rest on the right heel. The left leg is bent at the knee; the shin straight up and down. The left foot is flat on the deck. See figure 4-10.

Figure 4-10. Weaver High Kneeling.

- Medium kneeling: the right ankle is straight with the foot stretched out and the bootlaces in contact with the deck. The left leg is bent at the knee, the left foot flat on the deck. The right shin may be angled to the body to create a tripod of support for the position. See figure 4-11.

Figure 4-11. Weaver Medium Kneeling.

- Low kneeling: the right ankle is turned so the outside of the foot is in contact with the deck and the buttocks are in contact with the inside of the foot. The right shin may be angled to the body to create a tripod of support for the position. See figure 4-12.

Figure 4-12. Weaver Low Kneeling.

- Two-knee: drop both knees onto the deck. The toes may be curled to get into and out of the position quickly. Depending on the need for observation of the enemy, the buttocks may or may not rest on the heels. See figure 4-13.

Isosceles Kneeling Position

The advantage of the Isosceles variation is that it enables the Marine to fire over the top of cover while exposing less of the body to a threat. To assume the Isosceles kneeling position, the following steps are basic to all adaptations:

- Drop the right foot back or step forward with the left foot and place the right knee on the deck.
- Square the shoulders to the target.
- Extend the arms toward the target.
- Lean forward with the shoulders rolled forward and the head tucked between the shoulders to better manage recoil.

Figure 4-13. Weaver Two-Knee Kneeling.

The following steps provide specifics for each adaptation:

- High kneeling: the toes of the right foot are curled and in contact with the deck or the inside of the foot may be in contact with the deck. Depending on the need for observation, the buttocks may or may not rest on the right heel. The left leg is bent at the knee, the shin is straight up and down. The left foot is flat on the deck. See figure 4-14.

Figure 4-14. Isosceles High Kneeling.

- Medium kneeling: the right ankle is straight with the foot stretched out and the bootlaces in contact with the deck. The left leg is bent at the knee, the left foot is flat on the deck. The right shin may be angled to the body to create a tripod of support for the position. See figure 4-15.

Figure 4-15. Isosceles Medium Kneeling.

- Low kneeling: the right ankle is turned so the outside of the foot is in contact with the deck and the buttocks are in contact with the inside of the foot. The right shin may be angled to the body to create a tripod of support for the position. See figure 4-16.

Figure 4-16. Isosceles Low Kneeling.

- Two-knee: drop both knees onto the deck. The toes may be curled to get into and out of the position quickly. Depending on the need for observation of the enemy, the buttocks may or may not rest on the heels. See figure 4-17.

Figure 4-17. Isosceles Two-Knee.

Prone Position

The prone position is easily assumed, stable, and presents a small target to the enemy. Since the prone position places most of the body on the deck, it offers great stability for long range shooting. However, it is the least mobile of the firing positions and may restrict the field of view. The prone position can be adapted to either the Weaver variation or the Isosceles variation.

Weaver Prone Position

The Weaver variation of the prone position produces a cocked leg position by angling the body to the target and cocking the leg to support the position. The Weaver prone position is ideal for firing from behind cover. See figure 4-18.

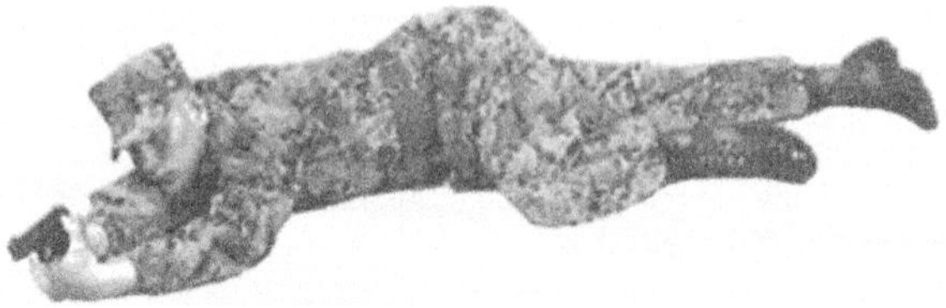

Figure 4-18. Weaver Prone Position.

To assume the Weaver prone position, perform the following steps:

- Face the target and make a half turn to the right (this places the body at a 40 to 60 degree oblique to the target). Grip the pistol in the right hand, placing the pistol in a position that facilitates control of the weapon. Ensure the pistol is pointed in a safe direction and does not cover any portion of the body.
- Move the body to the deck by using either the squat or drop method, keeping the body at a 40 to 60 degree oblique to the target.

  Squat Method
  - Squat down and place the left hand on the deck.
  - Kick both feet backward and come down on the right side of the body with the right arm extended toward the target.
  - Ensure the pistol does not cover the body or the left hand. See figure 4-19.

  Drop Method
  - Drop to a kneeling position.
  - Place the left hand on the deck in front of the body, push the pistol out toward the target.
  - Roll the right side of the body onto the deck.
  - Ensure the pistol does not cover the body or the left hand. See figure 4-20 on page 4-12.
- Bring the left knee up to support the firing position and to raise the diaphragm off the deck so as not to interfere with breathing. The inside of the knee rests on the deck. The knee is drawn up to provide maximum stability for the position.
- Establish a two handed-firing grip on the pistol.

Figure 4-19. Squat Method.

- Place the left elbow on the ground for stability. For maximum stability, strive to keep the grip firmly placed on the deck.
- The head may rest against the right arm so the pistol sights can be aligned. The head may be canted as long as the aiming eye can look directly through the sights. Strive to keep the pistol sights as level as possible while acquiring sight alignment.

To make minor increases in elevation, keep the left hand in place and firmly on the deck and raise the right hand to achieve the desired elevation. However, contact between the right and left hands must be maintained to stabilize the pistol. See figure 4-21 on page 4-12. (There is a tradeoff between obtaining the needed elevation and losing stability, so the Marine must strike a balance between the two.)

Figure 4-20. Drop Method.

Figure 4-21. Increasing Elevation (Minor Adjustments).

Isosceles Prone Position

The Isosceles variation of the prone position produces a straight leg position. See figure 4-22.

To assume the Isosceles prone position:

- Stand facing the target. Grip the pistol in the right hand, placing the pistol in a position that facilitates control of the weapon. Ensure the pistol is pointed in a safe direction and does not cover any portion of the body.
- Move the body to the deck by using either the squat or drop method.

  <u>Squat Method</u>
  - Squat down and place the left hand on the deck.
  - Kick both feet backward and come down on the right side of the body with the right arm extended toward the target.
  - Ensure the pistol does not cover the body or the left hand. See figure 4-23.

  <u>Drop Method</u>
  - Drop to a kneeling position.
  - Place the left hand on the deck in front of the body.

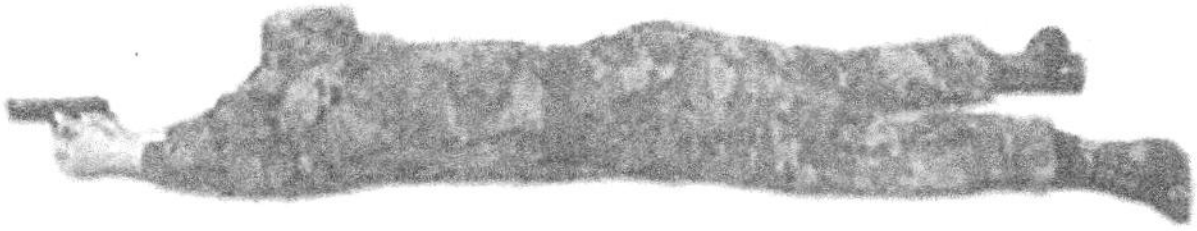

Figure 4-22. Isosceles Prone Position.

Figure 4-23. Squat Method.

  - Push the pistol out toward the target, and roll the right side of the body onto the deck.
  - Ensure the pistol does not cover the body or the left hand. See figures 4-24.
- Establish a two-handed firing grip on the pistol.
- Spread the legs to a position that provides maximum stability. The insteps of both feet may be flat on the deck or the toes may be curled and dug into the deck.
- Keep the pistol sights as level as possible while acquiring sight alignment. Keep the head in a position to allow the aiming eye to look directly through the sights.
- When wearing a helmet, the head may be canted slightly and rest against the right arm to push the helmet from the eyes so the sights can be aligned. Likewise, the pistol may be canted outboard to allow the aiming eye to look directly through the sights.

To make minor increases in elevation, keep the left hand in place and firmly on the deck and raise the right hand to achieve the desired elevation. However, contact between the right and left hands must be maintained to stabilize the pistol. See figure 4-25 on page 4-14. (There is a tradeoff between obtaining the needed elevation and losing stability, so the Marine must strike a balance between the two.)

Natural Body Alignment

The body must be properly aligned to the target so the sights fall naturally on the target when the pistol is presented. It takes a combination of body alignment and consistent muscular tension to

Figure 4-24. Drop Method.

Figure 4-25. Increasing Elevation (Minor Adjustments).

ensure the sights fall naturally to the same area of the target every time the pistol is presented. The Marine can execute the following to check natural body alignment and to ensure the sights center on the aiming area:

- Orient the body to a target and establish a variation of the standing position and a two-handed firing grip on the pistol. Aim in on the target.
- Close the eyes and take a deep breath.
- Open the eyes and see where the pistol sights are in relation to the target. If the pistol sights are right or left of the target, move the feet to adjust the position right or left.

  Note: Do not force the sights onto the target area by moving the arms; this increases the muscular tension on one side of the body, disturbs balance, and makes recoil harder to manage.

If the sights are significantly out of alignment when the pistol is at eye level, it may be an indication of a poor grip. When the grip is correct (to include the muscular tension in the grip, wrist, and forearms), the sights should align to the point that only minor adjustments are needed to align the sights to the aiming eye.

Repeat the preceding steps as necessary. Body alignment and muscular tension are correct when the sights are naturally placed in the same area on the target every time the Marine aims on the target.

CHAPTER 5
USE OF COVER AND CONCEALMENT

On the battlefield, a firing position that allows maximum observation of the enemy as well as cover and concealment is a necessity. A good position provides a solid foundation for the pistol, maximizes the use of cover to provide protection from enemy fire, allows mobility, and provides observation of the enemy. Where possible, the cover should be used to provide additional support for the position.

When contact with the enemy is made, it is important to seek cover as quickly as possible. Cover is anything that provides protection from enemy fire. Cover should be, at a minimum, thick enough to stop small arms fire and high enough to protect the Marine when firing from behind it. The effective use of cover enables engagement of enemy targets while affording protection from enemy fire. Cover can also be effectively used to conceal the Marine from enemy view while searching for targets.

Concealment is anything that hides a Marine from enemy view; however, it might not afford protection. Concealment can be provided by brush, trees, etc.

> *Note:* The procedures in this chapter are written for right-handed Marines; left-handed Marines must reverse directions as needed.

Cover Materials

Natural cover (rocks, logs, rubble, etc.) is best because it is hard to detect. But any material (including buildings, structures, etc.) that may protect an individual from small arms fire can be used for cover. Some common types of cover material are as follows.

Dirt

The best type of cover is dirt packed to a minimum thickness of 18 inches.

Cinder Blocks

Cinder blocks used as foundations for houses or walls can be used for cover, but they can be penetrated. They are also brittle and can shatter upon impact from small arms fire, which can cause injury by secondary fragmentation.

Trees, Logs, and Telephone Poles

Wood is a relatively dense material; therefore, it offers good covering protection since bullets have a tendency to fragment as they penetrate. Live trees have a greater resistance to bullet penetration than dead wood. Wood that has been treated with creosote, such as telephone poles and railroad ties, offers better protection from projectiles than untreated wood, but it still does not ensure protection from small arms fire.

Sandbags

Sandbags can be used for cover. However, they should always be packed tightly and tamped down to increase their density. If loosely filled or moist, a bullet can more easily penetrate the sandbag. Doubling or overlapping sandbags also increases their protective qualities.

Considerations for Firing From Cover

Adjusting the Firing Position

Cover can provide additional support for the firing position. The firing position is adjusted to ensure stability, mobility, and observation of the enemy.

The Marine chooses a firing position based on his height in relation to the height of the cover. The firing position must minimize exposure to the enemy but allow observation of the area.

Although the cover provides additional support, the Marine continues to apply the same amount of muscular tension in the grip, wrist, and forearms. Muscular tension is still necessary to stabilize the pistol sights and manage recovery.

Because the sights are higher than the muzzle of the pistol, the Marine must ensure that the muzzle of the pistol clears the cover as he obtains sight alignment/sight picture on the target. The closer the Marine is to the cover, the easier it is to ensure the muzzle clears the cover.

Applying the Weaver and Isosceles Variations

The type of cover may dictate which variation of a firing position will be the most effective. The position should provide the Marine with the maximum amount of stability and control and allow the Marine to manage recoil effectively in order to recover on target.

Weaver Variation

The Weaver variation exposes less of the body from behind cover due to the angle of the body. The Weaver variation may be better suited for firing from behind the right or left side of cover. For example, the Weaver prone is ideal for firing from behind a log.

Isosceles Variation

The Isosceles variation is good for firing over the top of cover (e.g., a window). The Isosceles prone is ideal for firing from behind narrow cover (e.g., a telephone pole).

Keeping the Body Behind Cover

Avoid inadvertent exposure of any part of the body. Be especially aware of the top of the head, elbows, knees, or any other body part that may extend beyond the cover.

Log or Curb

When firing from behind a log or curb, the Marine must present the lowest possible silhouette and may use the log or curb for maximum support of the position. For maximum protection, muzzle clearance is kept as close as possible to the top of the log or curb. The Marine may fire from either the side or the top of the log, depending on cover and concealment (see fig. 5-1). The Marine fires over the top of a curb when it is used for cover.

Figure 5-1. Firing Around a Log.

Wall or Barricade

Firing is done from either the side or over the top of a wall or barricade. See figure 5-2.

Figure 5-2. Firing From a Barricade.

Window

If the Marine has not been detected by the enemy, he should use the side of the window or the window sill for support. Ideally, it is best to fire from the corner of the window sill when using the window for support. If there is little chance for detection or the shot can be made without support, the Marine should remain back and to the side of the window opening so the pistol does not protrude and his body is concealed by the shadows/darkness of the room. If the Marine is positioned too close to the window, his body provides a silhouette to the enemy.

Vehicle

In many combat situations, particularly in urban environments, a vehicle may be the best form of cover. When using a vehicle for cover, the engine block provides the most protection from small arms fire. The Marine establishes a position behind the front wheel or front door jamb so the engine block is between him and the target. See figure 5-3.

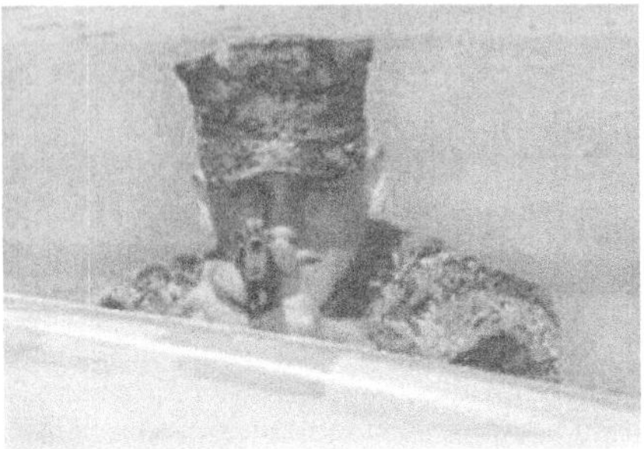

Figure 5-3. Firing Over the Hood of a Vehicle.

From this position, the Marine fires over the hood of the car or underneath the car from behind the wheel. At the back of the car the axle and the wheel provide the only cover. If the Marine must shoot from the back of the car, he must position himself directly behind the wheel as much as possible. See figure 5-4.

Figure 5-4. Firing From the Back of a Vehicle.

Providing Support for the Position and the Pistol

Support helps stabilize both the firing position and the pistol and enable the Marine to maintain sight alignment and sight picture.

The forearms or hands can contact the support to stabilize the pistol. The Marine may rest the pistol on or against the support as long as the support does not interfere or affect the pistol's cycle of operation. See figure 5-5.

Figure 5-5. Hand Resting on Support.

When firing over the top of cover, the Marine can establish a supported position and stabilize the position by resting the trigger guard or the magazine on the cover. The pistol may be pushed up against the support so the "V" formed by the receiver and the front of the trigger guard rests firmly against the support. See figure 5-6.

Figure 5-6. Trigger Guard Resting Against Support.

When firing around the sides of cover, the Marine can establish support and stabilize the position by placing the back of the hand or arm against the cover. Avoid placing the slide of the pistol against the cover because it can interfere with the pistol's cycle of operation. However, the pistol can be canted and placed against the cover so the trigger guard or the "V" formed between the receiver and trigger guard rests against the support. This position enables the Marine to expose less of himself to the enemy. See figure 5-7.

Figure 5-7. Hand Resting Against Side of Support.

When using a vehicle for cover, the Marine can establish additional support for the pistol by positioning himself in the car behind the door jamb (frame of door) and placing his hands or pistol against the "V" formed by the open door and door frame. See figure 5-8.

When shooting from the left side of cover, the Marine still uses his right hand and eye. See figure 5-9. He may have to cant his head and the pistol to the left to establish sight alignment. For right-handed Marines, shooting from the left side of cover may expose more of the Marine to the enemy than shooting from the right side.

Figure 5-8. Firing From Behind the Door Jamb of a Vehicle.

Figure 5-9. Firing From the Left Side of Cover.

Changing Positions

If the Marine has been firing from cover and has to reload or clear a stoppage behind cover he should attempt to resume firing from a different position. The enemy is aware of the Marine's current position and will be ready to engage him once he re-appears.

Moving

In combat, the Marine must be constantly aware of his surroundings and the available cover should enemy contact occur during movement from one position to another.

When moving from cover to cover, the Marine selects the next cover location and plans the route of movement before leaving his present position. This is done by quickly looking from behind cover to ensure the area is clear, ensuring the head and eyes are exposed for as short a time as possible. If necessary, the Marine should conduct a Condition 1 reload before moving from cover. Once the Marine is committed to moving, he must focus on the move until cover is reassumed.

Supported Firing Positions

During combat, the Marine may not have the time to assume a perfect firing position. He must know instinctively that his position is correct rather than follow a regimented sequence of movements to ensure its correctness. With training, the Marine can assume stable firing positions quickly and instinctively by incorporating the use of cover for support. Support provides foundations for the firing position; which, in turn, provides support for the pistol. To maximize the support provided by the position, the firing position should be adjusted to fit or conform to the shape of the cover.

A supported firing position minimizes exposure to the enemy, maximizes the stability of the pistol and protection from fire, and provides observation of the enemy. Any stable support may be used (e.g., logs, sandbags, walls). The surrounding environment dictates the support and position.

The size, distance to the target, and time affect the need for stability and recovery in the selection of a supported firing position. For example, if the target is a great distance from the Marine, he may sacrifice some of his ability to manage recoil in order to assume a supported firing position that provides him the additional stability needed to fire accurately at long range. But, recovery may be more important for the Marine if he must fire multiple shots on target quickly; therefore, he may sacrifice some stability in his supported firing position in order to engage a target with multiple shots.

Supported Prone

The supported prone position presents the lowest silhouette and provides maximum protection from enemy fire. The supported prone position can be assumed behind a tree, a wall, or almost any type and size of cover. It is flexible and allows firing of the pistol from all sides. To assume the supported prone position and maximize the use of cover, the position is kept as low as possible to ensure no part of the body is exposed to the enemy. If the cover is narrow, the Marine positions his body directly behind cover and keeps his legs together. The Marine's body is in line with the pistol and directly behind the pistol. The Isosceles prone position presents a smaller target to the enemy and more body mass to absorb recoil. The Weaver prone position is ideally suited to fire from around cover (e.g., log) because of the angle of the body, but this position presents a larger target to the enemy. See figure 5-10.

Figure 5-10. Weaver Prone.

Supported Kneeling

If the prone position cannot be used because of the height of the support, the Marine may use the supported kneeling position. The kneeling position allows firing of the pistol from all sides. This position may be altered to maximize the use of cover or support by assuming a variation of the kneeling position (high, medium, low, or two-knee). The kneeling position provides more mobility than the prone position. See figure 5-11.

Caution

The Marine must not indicate to the enemy his position with his exposed knee. If the Marine changes knees while exposed to the enemy, he has indicated the direction of his next shot to the enemy.

Figure 5-11. Supported Kneeling.

Supported Standing

When use of the support is maximized, the supported standing position can be as stable as the supported kneeling or prone position. The supported standing position provides greater mobility than the other positions and usually provides greater observation of the enemy. In the standing position, the Marine must not allow the placement of his foot to indicate to the enemy his position behind cover. See figure 5-12.

Figure 5-12. Supported Standing.

Locate and Engage Targets From Behind Cover

To locate targets when behind cover or to ensure the area is clear before moving, the Marine must expose himself to the enemy. There are two techniques used to locate and engage targets from behind cover: the pie and the rollout. These techniques minimize the Marine's exposure to enemy fire while placing him in a position to engage targets or to move to another location if necessary. Both techniques are used in the kneeling and standing positions.

Pie Technique

The Weaver position is the most effective position when executing the pie technique because the position of the Marine's body minimizes exposure to the enemy. See figure 5-13.

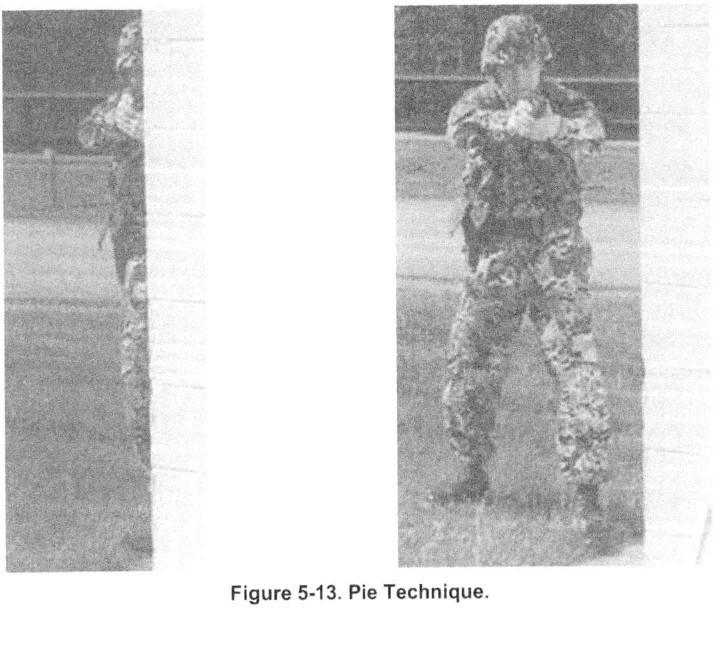

Figure 5-13. Pie Technique.

To perform the pie technique—

- Stay behind cover while moving back and away from the leading edge of the cover. The surroundings and situation dictate the distance the Marine moves back and away from the cover. Generally, the further back the Marine is from cover, the greater his area of observation; staying too close to cover decreases the area of observation.
- Assume a firing position and lower the pistol sights enough to have a clear field of view, aiming in on the leading edge of the cover.
- Take small side steps and slowly move out from behind cover, covering the field of view with the aiming eye and muzzle of the pistol. Wherever the eyes move, the muzzle should move (eyes, muzzle, target).
- Continue moving out from cover until a target is identified or the area is found to be clear. If a target is identified, sweep the safety off, place the finger on the trigger, and engage the target.

Note: If a target is identified before moving out from cover, the pistol should be thumb-cocked and off safe before moving out.

Rollout Technique

In this technique, the Isosceles position is the most effective position when executing the rollout technique because the position of the body allows the Marine to better maintain his balance. See figure 5-14. To perform the rollout technique—

- Stay behind cover, move back, and position the body so it is in line with the leading edge of the cover, ensuring that no part of the body extends beyond cover.
- Assume firing position and come to the Ready, ensuring the muzzle is just behind cover.
- Cant the head and pistol slightly and roll the upper body out to the side enough to obtain a clear field of view. Keep the feet in place and push up on the ball of one foot to facilitate rolling out.
- Continue to roll out from cover until a target is identified or the area is found to be clear. If a target is identified, sweep the safety off, place the finger on the trigger, and engage the target.

Figure 5-14. Rollout Technique.

Note: If a target is identified before moving out from cover, the pistol should be thumb-cocked and off safe before moving out.

Combining the Pie and Rollout Techniques

In some situations, it may be necessary to utilize both the pie and rollout technique in order to search an entire area for targets (i.e., corner of a building, doorway). Changing from one technique to another may allow the Marine to minimize his exposure to the enemy and reduce the time he is exposed to enemy fire.

Chapter 6
Presentation of the M9 Service Pistol

In combat, targets present themselves with little or no warning. The Marine must have the ability to react quickly and to effectively and efficiently present the pistol, whether the pistol is in the holster or at a carry. To successfully engage a combat target with the M9 service pistol, the Marine must master pistol presentation from the carries and transport, pistol presentation while assuming a firing position, and search and assess techniques.

Note: The procedures in this chapter are written for right-handed Marines; left-handed Marines should reverse directions as needed.

Sight Alignment/Sight Picture

Pistol presentation drills and dry fire help the Marine achieve a consistent grip and rapid presentation and aid in quickly acquiring sight alignment and sight picture. The Marine always executes the following steps:

- Identify the target and quickly present the pistol to the target while simultaneously sweeping the safety with the thumb of the right hand. (Disengaging the safety with the right thumb ensures the trigger is not pulled before taking the pistol off safe.)
- Shift the focus from the target to the front sight to obtain sight alignment while presenting the pistol. As the front sight breaks the plane of vision, acquire the front sight and begin to apply trigger control as sight picture is acquired.
- Apply trigger pressure until the shot is fired.

Presentation From the Carries and Holsters

Presentation From the Ready

To present the pistol from the Ready, the Marine performs the following steps in sequence. When a target appears—

- Sweep the safety with the thumb of the right hand, place the trigger finger on the trigger, and raise the arms to bring the pistol to the target.

  Note: If the Marine wishes to thumbcock the pistol for a single action shot, the pistol is thumbcocked with the left thumb after the safety is swept with the right thumb. The grip of the left hand may have to be broken to thumbcock the pistol; re-establish the grip after thumbcocking.

- Acquire sight alignment and sight picture within the aiming area and apply trigger pressure until the shot is fired.

Presentation From the Alert

To present the pistol from the Alert, the Marine performs the following steps in sequence. When a target appears—

- Sweep the safety with the thumb of the right hand, place the trigger finger on the trigger, and bring the pistol to the target:
  - If the arms are straight, raise the arms to a 45-degree angle with the deck.
  - If the arms are bent, straighten the arms out toward the target.

Note: If the Marine wishes to thumbcock the pistol for a single action shot, the pistol is thumbcocked with the left thumb after the safety is swept with the right thumb. The grip of the left hand may have to be broken to thumbcock the pistol; re-establish the grip after thumbcocking.

- Acquire sight alignment and sight picture within the aiming area and apply trigger pressure until the shot is fired.

Presentation From a Holster Transport

The pistol is presented from the holster in one continuous, fluid motion.

M12 Holster

To present the pistol from the M12 holster transport, perform the following steps in sequence once the target appears:

- Place the heel of the left hand at the center of the torso with the fingers extended toward the target. (The placement of the left hand allows a two-handed grip to be established in a minimum amount of movement.) At the same time, unfasten and release the D-ring with the right hand.
- Use the right hand to place the thumb on the forward edge of the holster and the fingers around the back edge of the holster, keeping the trigger finger straight.
- Slide the hand up the holster until the fingers come in contact with the pistol grip. At the same time, keep the thumb above the pistol to guide the holster flap up.
- Grasp the pistol grip with the fingers and draw the pistol straight up. Continue withdrawing the pistol while moving the thumb to a position on the safety.

Note: Hand placement should allow the firing grip to be established once the thumb disengages the safety. Any adjustments made to the firing grip after the safety has been disengaged should be minor.

- Once the muzzle clears the holster, rotate the muzzle forward while sweeping the safety.
- Establish a two-handed grip on the pistol by joining the left hand with the right hand in front of the torso. At the same time, start to straighten the pistol out toward the target. See figure 6-1.

Figure 6-1. Presentation From the M12 Holster—Step 1.

- Continue moving the pistol toward the target and, at the same time, place the trigger finger on the trigger, acquire sight alignment and sight picture within the aiming area, and apply trigger pressure until the shot is fired. See figure 6-2.

Figure 6-2. Presentation From the M12 Holster—Step 2.

When time permits to thumbcock the pistol for a single action shot (e.g., a long-range engagement), the Marine presents the pistol from the holster by performing the following steps once the target appears:

- Withdraw the pistol from the holster.
- Sweep the safety with the thumb of the right hand.
- Join the left and right hands and thumbcock the pistol with the left thumb.
- Establish a two-handed grip on the pistol.
- Straighten the pistol out toward the target and, at the same time, place the trigger finger on the trigger, acquire sight alignment and sight picture within the aiming area, and apply trigger pressure until the shot is fired.

M7 Shoulder Holster

The M9 service pistol fits very snugly in the M7 shoulder holster because the holster was originally designed for the M1911A1 .45-caliber pistol, which has a more slim, round design, particularly around the trigger guard. A firm grip is required when holstering and withdrawing the M9 service pistol from the M7 shoulder holster. To present the pistol from the M7 shoulder holster, perform the following steps:

- Unsnap the thumb snap closure with the left hand.
- Wrap the fingers of the right hand around the pistol grip and rest the thumb on top of the inside of the holster. See figure 6-3.

Figure 6-3. Presentation From the M7 Shoulder Holster—Step 1.

- Grasp the pistol grip firmly and draw the pistol up and away from the holster while rotating the thumb in a position to operate the safety.
- Continue withdrawing the pistol until the muzzle clears the holster and rotate the muzzle toward the target. See figure 6-4.

Figure 6-4. Presentation From the M7 Shoulder Holster—Step 2.

- Sweep the safety with the thumb of the right hand while starting to punch the pistol out toward the target.
- Establish a two-handed grip on the pistol by joining the right hand with the left hand in the center of the torso. See figure 6-5.
- Continue punching the pistol out and, at the same time, place the trigger finger on the trigger, establish sight alignment and sight picture within the aiming area, and continue trigger pressure until the shot is fired.

Assault Holster

To present the pistol from the assault holster, perform the following in one, continuous motion:

Figure 6-5. Presentation From the M7 Shoulder Holster—Step 3.

- Bring the heel of the hand down on the hammer of the pistol so that it disengages the thumb break. See figure 6-6.
- Sweep the hand forward in a small circular motion, slide the hand up the holster, and bring the fingers up under the pistol grip.

Note: Depending on the type of retention strap, this circular motion releases the retention strap or pushes the retention strap forward and out of the way.

Figure 6-6. Presentation From the Assault Holster—Step 1.

- Grasp the pistol grip and draw the pistol straight up and out of the holster while establishing a firing grip and positioning the thumb on the safety to operate it. See figure 6-7.

Figure 6-7. Presentation From the Assault Holster—Step 2.

Concealed Pistol Holster

Depending on the clothing worn by the Marine, there are two methods for accessing the concealed pistol holster so that the clothing does not become an obstruction while presenting the pistol. To present the pistol from the concealed pistol holster, the following steps are performed in one, continuous motion when the target appears:

- If the Marine is wearing a short jacket/sweater (waist-level) that is buttoned or zipped—
  - Move the left hand across the body and grasp the jacket or sweater just above the holster, at the same time place the right hand on the holster below the jacket or sweater. See figure 6-8 on page 6-6.
  - Pull up on the jacket or sweater with the left hand. At the same time, slide the right hand up the holster until the fingers come in contact with the pistol grip. Keep the thumb above the pistol to ensure the jacket or sweater clears the pistol. See figure 6-9 on page 6-6.

Note: Leaning slightly forward at the waist may assist in clearing the jacket or sweater from the pistol and in removing the pistol from the holster.

- If the Marine is wearing a long jacket (below waist-level), or a short jacket or sweater that is unfastened—
  - Place the heel of the left hand at the center of the torso to hold the left side of the jacket down with the fingers extended. At the same time, place the right hand on the right side of the jacket, with the thumb underneath the right side of the jacket below the left hand. See figure 6-10 on page 6-7.
  - Keep thumb of the right hand against the body and throw the jacket back and away from the holster. See figure 6-11 on page 6-7.

Note: Keeping a weighted object (e.g., keys, extra magazine) in the right-hand pocket of the jacket assists in throwing the jacket back and away from the holster.

  - Place the right hand on the holster and slide the hand up the holster until the fingers come in contact with the pistol grip.
- Grasp the pistol grip with the fingers and draw the pistol straight up. Continue withdrawing the pistol while moving the thumb to a position on the safety.

Note: Hand placement should allow the firing grip to be established once the thumb disengages the safety. Any adjustments made to the firing grip after the safety is disengaged should be minor.

- Rotate the muzzle forward, once it clears the holster, while sweeping the safety.
- Establish a two-handed grip on the pistol by joining the left hand with the right hand in front of the torso. At the same time, start to punch the pistol out toward the target.

Figure 6-8. Presentation From the Concealed Pistol Holster—Step 1.

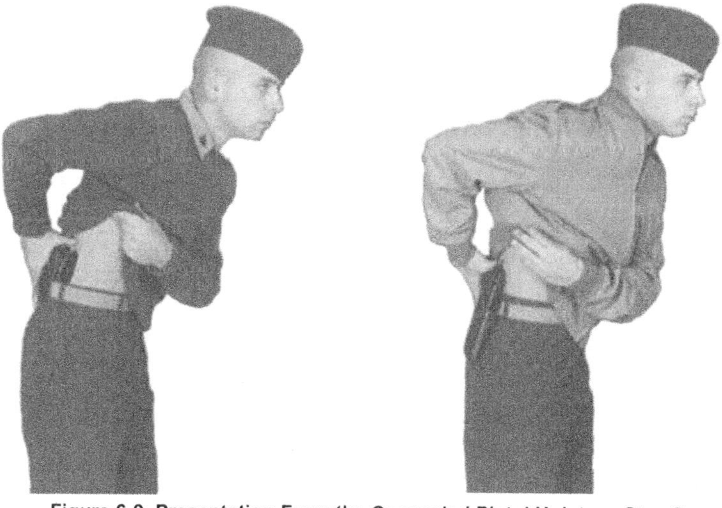

Figure 6-9. Presentation From the Concealed Pistol Holster—Step 2.

Figure 6-10. Presentation From the Concealed Pistol Holster with the Long Jacket—Step 1.

Figure 6-11. Presentation From the Concealed Pistol Holster with the Long Jacket—Step 2.

- Continue punching the pistol out and, at the same time, place the trigger finger on the trigger, acquire sight alignment and sight picture within the aiming area, and apply trigger pressure until the shot is fired.

Presentation While Assuming the Kneeling Position

In combat, the Marine must be able to quickly assume the kneeling position while presenting the pistol to the target. This usually takes advantage of cover and provides a more stable base for shooting. The following procedures are based on the Isosceles and Weaver variations of the standing to the kneeling shooting positions.

From the Ready

The Marine drops to the kneeling position while raising the arms to bring the pistol up to the target.

Note: The Weaver variation maintains the body's position in a 40 to 60 degree oblique to the target.

At the same time, the Marine sweeps the safety with the thumb of the right hand, places the trigger finger on the trigger, acquires sight alignment and sight picture within the aiming area, and applies trigger pressure until the shot is fired.

From the Alert (Arms Straight)

The Marine drops to the kneeling position while raising the arms to bring the pistol up to the target.

Note: The Weaver variation maintains the body's position in a 40 to 60 degree oblique to the target.

At the same time, the Marine sweeps the safety with the thumb of the right hand, places the trigger finger on the trigger, acquires sight alignment and sight picture within the aiming area, and applies trigger pressure until the shot is fired.

From the Alert (Close Quarters, Elbows Bent)

The Marine drops to the kneeling position while pushing the arms out toward the target.

> *Note:* The Weaver variation maintains the body's position in a 40 to 60 degree oblique to the target.

At the same time, the Marine sweeps the safety with the thumb of the right hand, places the trigger finger on the trigger, acquires sight alignment and sight picture within the aiming area, and applies trigger pressure until the shot is fired.

From the Holster Transport

The Marine withdraws the pistol from the holster while dropping to the kneeling position. The pistol should be rotated to the target by the time the knee hits the deck. The Marine sweeps the safety with the thumb of the right hand, establishes a two-handed firing grip and—

- In the Weaver variation, raises the arms to bring the pistol up toward the target.
- In the Isosceles variation, pushes the pistol out toward the target.

At the same time, the Marine places the trigger finger on the trigger, acquires sight alignment and sight picture within the aiming area, and applies trigger pressure until the shot is fired.

Presentation While Assuming the Prone Position

The prone position is generally assumed to take advantage of cover or to provide additional stability for shooting. The following procedures are executed from the standing holster transports.

Isosceles Prone

The Marine withdraws the pistol from holster. At the same time, the Marine eliminates body contact with the deck by either the squat method or the drop method. The pistol is rotated to the target as the left hand is placed on the deck. The Marine sweeps the safety off.

> *Note:* If the Marine wishes to thumbcock the pistol for a single action shot, it is done once the Marine is on the deck.

The Marine establishes a two-handed firing grip while spreading the legs a comfortable distance apart for stability. He places the trigger finger on the trigger, acquires sight alignment and sight picture within the aiming area, and applies trigger pressure until the shot is fired.

Weaver Prone

The Marine withdraws the pistol from the holster. At the same time, the Marine eliminates body contact with the deck by either the squat method or the drop method. The Marine maintains the 40 to 60 degree oblique to the target. The pistol should be rotated to the target as the left hand is placed on the deck. The Marine sweeps the safety off.

> *Note:* If the Marine wishes to thumbcock the pistol for a single action shot, it is done once the Marine is on the deck.

The Marine brings the left knee up, establishes a two-handed firing grip, and places the left elbow on the deck for stability. He places the trigger finger on the trigger, acquires sight alignment and sight picture within the aiming area, and applies trigger pressure until the shot is fired.

Search and Assess

To be successful in combat, the Marine must have the ability to assess the situation and take appropriate action following engagement. Once the Marine fires one or two rounds to engage a target, he must quickly assess the situation and the effectiveness of his engagement. This split-second assessment allows the Marine to determine the best course of action and prevents him from expending a number of rounds without assessing the results and his ability to accurately engage the target.

To search and assess, perform the following steps in sequence:

- Place the trigger finger straight along the receiver immediately after a target is engaged. Do not place the pistol on safe.
- Lower the arms just enough to look over the pistol sights and provide a clear field of view. Do not rotate the wrists in order to angle the pistol downward, this breaks the firing grip and changes the tension in the arms. Lowering the arms maintains the firing grip, keeps the sights level, and allows sight alignment and sight picture to be quickly re-established should follow-on shots need to be fired.
- Search the area by moving the head, eyes, and pistol left and right (approximately 45 degrees from center). Wherever the head moves, the muzzle moves (eyes, muzzle, target). Keeping both eyes open increases the field of view.

If the target has not been eliminated, the Marine must determine whether to re-engage the target:

- If the Marine decides to re-engage the target, the target's size, time and distance to the target, and Marine's capabilities dictate the technique used. Re-engagement techniques include re-engaging the target with a well-aimed precision shot, two shots, or offset aiming.
- If the target's distance exceeds the Marine's engagement capabilities or if the target is partially exposed and too small for the Marine to accurately engage, the Marine should not attempt to re-engage the target. The Marine may seek cover, seek out a better opportunity for engagement, or use support to better stabilize the pistol.

When it is determined that the area is clear of all enemy threat, place the pistol on safe without breaking the grip and assume a pistol carry or pistol transport.

CHAPTER 7
PISTOL ENGAGEMENT TECHNIQUES

The Marine must be able to quickly detect targets and employ the proper pistol engagement and firing techniques. The Marine must also be able to re-engage a target if the initial engagement is not successful.

Note: The procedures in this chapter are written for right-handed Marines; left-handed Marines should reverse directions as needed.

Target Detection

Target indicators reveal an enemy's position. Most combat targets are detected by smoke, flash, dust, noise, or movement and are usually only momentarily visible. Target indicators are grouped into three general areas: movement, sound, and improper camouflage.

Movement

The Marine does not need to look directly at an object to notice movement. The eye is attracted to any movement, especially sudden movement. A slowly moving target is harder to detect than one with quick, jerky movements. Therefore, the ability to locate a moving target depends primarily on the speed of the object's movement.

Sound

Sound (e.g., movement, rattling equipment, talking) can be used to detect an enemy position. Sound provides only a general location, making it difficult to pinpoint a target by sound alone. However, sound can alert the Marine to the presence of a target and increases the probability of locating the target through other indicators.

Improper Camouflage

The improper use of camouflage creates three target indicators: shine, outline, and contrast with the background. The closeness of a typical pistol engagement somewhat diminishes these effects, but can still affect target detection. The Marine's ability to recognize target indicators aids in the detection of targets.

Shine

Shine is created from reflective objects such as metal, glass, pools of water, and the skin's natural oils. Shine acts as a beacon in locating a target's position.

Outline

The outline of camouflaged objects such as the body, head and shoulders, weapons, and gear can be recognizable.

Contrast With the Background

Objects contrast with a background because of differences in color, surface, and shape. The following are examples of objects contrasting with their background:

- A target wearing a dark uniform would be clearly visible in an area of snow or sand.
- Symmetrical shapes, such as helmets or rifle barrels, can be detected in a wooded area.
- Fresh soil around a fighting hole contrasts with the otherwise unbroken ground surface.

While observing an area, the Marine notes anything that looks out of place or unusual and studies it in more detail in order to increase the chances of spotting a hidden threat.

Techniques of Fire

To successfully engage a combat target with the M9 service pistol, the Marine must employ effective techniques of fire. The Marine's performance of these skills and the proper application of the fundamentals of marksmanship are critical to success in a combat situation.

Double and Single Action Firing

When the M9 service pistol is taken off safe, it is capable of firing in both a double and a single action mode.

Double Action Mode

The design of the M9 service pistol causes the first shot fired to be a double action shot.

In double action firing, two actions occur as the trigger is moved to the rear; the hammer moves to the rear, cocking the pistol, and then the hammer moves forward, firing the pistol. More pressure is required on the trigger to fire a double action shot due to the distance between the trigger and hammer and the weight of the trigger. A double action shot requires approximately 9 to 16 pounds of pressure to move the trigger rearward.

Maintaining sight alignment and sight picture are harder when firing a double action shot. Therefore, the sights will more than likely move outside the aiming area when applying trigger pressure.

Single Action Mode

In single action firing, the pistol is already cocked because the cycle of operation in the first double action shot leaves the hammer cocked to the rear. Therefore, the only action taking place as the trigger is moved to the rear is the hammer moving forward, firing the pistol. A single action shot requires approximately 4 to 6 pounds of pressure to move the trigger rearward. Therefore, the application of trigger control is easier when firing a single action shot.

To enable the first shot to be fired single action, the pistol's hammer can be manually cocked with the thumb:

> *Note:* The pistol must be taken off safe before it can be thumbcocked.

- Use the left thumb to pull back on the hammer to cock it. This ensures that the firing grip of the right hand does not have to be broken.
- Ensure that the hammer moves all the way to the rear.
- Ensure the trigger finger remains straight along the receiver until the pistol is fully cocked.
- Re-establish the firing grip with both hands once the pistol is cocked.

Single Action or Double Action Firing Factors

The decision to fire in a single or double action mode is made rapidly. Ultimately, the decision is based on the Marine's abilities, but the decision is also based on time and accuracy. The pistol is fired in the double action mode when trigger control, sight picture, and stability of hold are not as critical for accuracy (i.e., close range, large targets). The pistol is fired in the single action mode when the fundamentals are more critical to accuracy (i.e., long range, small targets).

Time

The Marine sacrifices time to fire a single action, precision shot; but what he sacrifices in time, he gains in accuracy.

For quick engagements at close range, there may not be time to thumbcock the pistol for a single action shot. Therefore, firing the first shot in the double action mode is preferred because shots are needed on target quickly and stability of hold and sight picture are not as critical to accuracy.

When time permits and for targets at longer ranges, the pistol may be thumbcocked to place it in the single action mode to reduce the weight of the trigger and the distance the trigger must travel rearward to fire the first shot.

Distance and Size of the Target

The smaller the target, the more critical the application of the fundamentals in order to engage the target accurately. To accurately engage a small target (e.g., head shot, long-range target) it is better to thumbcock the pistol for a single action shot. Engagement of a smaller target requires additional precision because sight alignment and sight picture are more critical to accuracy.

Two-Shot Technique

During combat, the preferred pistol engagement technique is to rapidly fire more than one shot on a target to eliminate it as a threat. Two shots fired in rapid succession increase the trauma (i.e., shock, blood loss) on the target and increase the Marine's chances of quickly eliminating the threat. Therefore, two shots are most often fired in rapid succession on a target at close range.

After the pistol is fired, the muzzle climbs with the recoil of the pistol. To fire two shots, the Marine must quickly recover the sights to the same area on the target while re-acquiring sight alignment and sight picture. The proper recovery automatically brings the sights back on target following recoil. A quick recovery allows more time for the Marine to align the sights and apply trigger control to fire the next shot. Recovery begins immediately after the application of the fundamentals to bring the pistol sights into alignment with the target in preparation for firing the next shot.

The amount the muzzle climbs during recoil depends on the amount of controlled muscular tension in the grip and wrists used to stabilize the pistol and create consistency in resistance to recoil. Controlled muscular tension allows the pistol sights to recover consistently back on target within a minimum amount of time. The speed of delivery of multiple shots depends on how fast the Marine can re-acquire sight alignment. Sight recovery is determined by the following:

- The key to proper recovery is a stable firing position and proper grip.
- If the Marine's firing position is not stable, recoil forces him out of his firing position, requiring him to re-establish his position before he takes his next shot.
- An improper grip or lack of controlled muscular tension causes the pistol to move in the Marine's hand after the shot is fired, disrupting sight alignment and requiring the Marine to re-establish his grip.

Slow Fire Technique

Sight alignment becomes more critical the smaller the target and the greater the distance to the target. In these situations, the Marine does not engage the target with two rapidly fired shots because he has to slow down his application of the fundamentals in order to fire a precision shot(s). To engage small targets (i.e., partially exposed) and targets at longer ranges where precision is required, the Marine must employ the following slow fire technique:

- Thumbcock the pistol for a single action shot.
- Slow down application of the fundamentals.
- Fire one well-aimed, precision shot on target.

Re-engagement Techniques

Once the Marine has quickly assessed the situation and determined that the threat still exists, he may make the decision to re-engage the target to eliminate it as a threat. Re-engagement techniques include engaging the target with a precision shot through slow fire, two shots, or offset aiming. The time, size, and distance to the target as well as the Marine's marksmanship abilities dictate the technique used to re-engage the target.

Slow Fire Technique

If the Marine has engaged a target and the target still poses a threat, the Marine may choose to slow down his application of the fundamentals and fire a slow fire, precision shot. The placement of one well-aimed, precision shot on a designated area of the target increases the chances of eliminating the target as a threat. However, the time the Marine has to engage the target determines if he can slow down his application of the fundamentals to fire a precision shot. Firing a precision shot takes time, so the Marine will sacrifice time for accuracy.

The key to successful shot placement is the Marine's ability to slow down his performance and focus on the application of the fundamentals of marksmanship. The Marine must determine where to place shot for maximum effectiveness. In order to make this decision, the Marine must consider distance and size of the target and time. For targets at close range and within the Marine's marksmanship abilities, a precision shot is placed in the head to immediately eliminate the target as a threat. Sometimes, even at close ranges, the target may only be partially exposed; therefore, the Marine is presented with a small target and aims his sights on the portion of the target that is exposed. For targets at long ranges, the target is smaller and requires a precision shot. A precision shot may be placed in the body to add trauma to the target and increase the chances of eliminating the target as a threat.

Two-Shot Technique

If a target is accurately engaged, but does not go down, the Marine may re-engage the target with additional shots. Two additional, rapidly delivered shots increases the trauma and increases the Marine's chances of eliminating the threat.

The size and distance to the target affect how quickly two shots can be delivered on the target. The speed at which two shots are fired also depends on the Marine's marksmanship abilities and how fast he can re-acquire his front sight. However, the Marine must not compromise accuracy for speed; the key to successful target engagement is to fire only as quickly as the Marine can fire effectively.

Offset Aiming

When the Marine assesses the situation and determines his shots are not successful and not striking the target in the designated aiming area, he may employ offset aiming. Since the pistol's sights cannot be adjusted, offset aiming is applied to adjust the aiming area and to cause rounds to strike center mass. This technique should only be applied when the Marine determines that he is applying the fundamentals of marksmanship correctly. Sight picture will change as the aiming area is adjusted.

The known strike of the round offset aiming technique requires shifting the point of aim to compensate for rounds striking off target center. To effectively engage a target using this technique, the Marine must be able to see where the rounds are striking and then aim an equal distance from the center of the aiming area opposite the observed strike of the round. For example, if the rounds are striking the target high and left of center mass, aim an equal and opposite distance low and right.

Multiple Targets

If engaging more than one target at a time, the Marine adheres to the fundamentals of marksmanship and employs the techniques of multiple target engagement. The introduction of multiple targets in a combat scenario requires additional skills that must be learned and practiced if a Marine is to be successful. To be effective in combat, the Marine must be able to detect targets through identification of target indicators (identified on page 7-1), prioritize the targets, and employ multiple target engagement techniques.

Combat Mindset

Successful engagement of multiple targets requires a somewhat different mindset from single target engagement. For example, following engagement of a single target, the Marine assesses the situation. During multiple target engagements, after the first target is engaged, he must immediately engage the next target and continue until all targets have been eliminated. Because split-second decisions must be made, the development of a combat mindset is important to success on the battlefield. A combat mindset allows the Marine to control the pace of the battle rather than react to the threat.

Mental preparedness is essential to successful engagement of multiple targets, and the required mindset must be developed until it becomes second nature to the Marine. When multiple targets appear, the Marine must prioritize the targets to establish an engagement sequence. To prioritize targets and establish an engagement sequence, the Marine must be aware of the surroundings, not focus on one target, and continuously search the terrain for additional targets.

Prioritizing Targets

Once multiple targets have been identified, they must be prioritized in terms of the threat each target presents. While the fundamentals of marksmanship must still be applied, prioritizing targets and planning the engagement are just as essential to successful multiple target engagement.

Target priority is based on factors such as proximity, threat, and opportunity. It also encompasses the Marine's proficiency level since a Marine should not attempt to engage a target beyond his proficiency level or the pistol's capability (e.g., a target 100 yards away). The principal method of prioritizing targets is to determine the level of threat for each target so that all targets may be engaged in succession, starting with the most threatening and ending with the least threatening.

Prioritizing targets is an ongoing process. As the engagement proceeds, new targets may appear or previously identified targets may take cover, delaying their engagement. Therefore, the Marine must remain constantly alert to changes in target threat, proximity, and the opportunity for engagement and revise target priorities accordingly.

Engaging Multiple Targets

Once targets have been prioritized, the Marine must quickly eliminate them. It is crucial to understand and practice the physical techniques for engaging multiple targets until they become second nature. When these techniques occur as automatic responses, the Marine maintains an awareness of the complete battlefield and is able to concentrate on the mental aspects of multiple target engagement.

Engagement Methods

The fundamentals of marksmanship are critical to the development of skills to support multiple target engagement. While responsiveness is important when engaging multiple targets, the Marine's primary concern is to place accurate rounds on target. A balance must be struck between placing shots quickly on targets at close range and slowing down to place precision shots on targets at long range because all targets pose a threat and must be accurately engaged. To engage multiple targets—

- Engage every target, move from target to target.
- After all of the targets have been initially engaged, assess the effectiveness of the engagements and, if necessary, re-engage targets that still pose a threat.
- If all of the targets are eliminated, search the area for new targets.

Considerations for Field Firing Positions

The selection and use of field firing positions are critical to engagement of multiple targets. As in any firing situation, if the situation permits, the Marine should make a quick mental review of the terrain to select a firing position that provides stability, mobility, and observation of the enemy. The firing position should also provide flexibility for engaging multiple targets. The more dispersed the multiple targets, the greater the lateral movement the position must afford to engage them.

The prone position provides the maximum stability for firing well-placed shots, but engaging multiple targets from this position may require adjustment in the position from shot to shot depending on the distance between the targets and their location. Because the arms are fully extended on the ground, the prone position can be restrictive, increasing the time it will take to recover the sights onto subsequent targets. If the Marine must make an adjustment in the prone position to engage multiple targets, he must ensure that he maintains proper arm placement. Improper arm placement may affect his stability of hold and his ability to acquire sight picture. The following identifies position adjustments:

- For targets that are not widely dispersed, the Marine adjusts the position of the lower part of the body to orient the upper body toward the target without disturbing the placement of the arms. This allows the Marine to maintain stability of hold and quickly recover on subsequent targets. Moving the legs to the left orients the upper body to the right; moving the legs to the right orients the upper body to the left.
- For targets that are widely dispersed, the Marine may need to adjust his entire position to orient his body in the direction of subsequent targets. Using the left hand, push the upper body off the deck just high enough to move the body in the direction of the target. Extend the right arm toward the target and keep the muzzle pointed in the direction of the target.

The kneeling and standing positions provide an increased field of view and allow maximum lateral adjustment to engage dispersed targets. The type of adjustment the Marine must make in the kneeling or standing position in order to orient his body in the direction of each target is related to the distance between the targets. It is also related to the Marine's ability to maintain stability of hold and recover from recoil. Once the first target is engaged—

- Maintain a proper shooting position, rotate the body in the direction of the target, keeping the feet in place. In the standing position, the Marine may bend slightly at the knees while rotating the body. This allows the Marine to distribute his weight forward to provide additional stability to the position and to better control the effects of recoil.
- Adjust the entire position by either adjusting the placement of the feet or knee(s) in order to face in the direction of the new target if the targets are some distance apart.
- Ensure that as the eyes move to a new target, the head and pistol muzzle should follow.

Moving Targets

The majority of combat targets will be moving; the enemy will move quickly from cover to cover, exposing himself for the shortest time possible. Therefore, a moving target must be engaged before it disappears. The Marine must engage a moving target with the same accuracy and precision used to engage a stationary target. Practice in the proper application of leads and the fundamentals of marksmanship enable the successful engagement of moving targets.

Types of Moving Targets

Moving Target

This type of target moves in a consistent manner (i.e., walking or running) and is in continuous sight as it moves across the Marine's field of vision. However, unless the enemy is completely unaware of the Marine's presence, this type of target is not likely to present itself.

Stop and Go Target

This type of target appears and disappears during its movement due to intermittent cover. It will present itself for only a short period of time before re-establishing cover. This type of target is most vulnerable to fire at the beginning and end of its movement because it begins slow and gains momentum as it leaves cover and then slows to avoid overrunning the new cover.

Leads for Moving Targets

Once the moving target has been identified it must be engaged. To engage a moving target, the Marine must aim at a point in front of the target, which is known as a lead. Lead is the distance in advance of the target that is required to strike the target when it is moving. To achieve success, determining the amount of lead to engage a moving target must be as precise as possible. When a shot is fired at a moving target, the target continues to move during the time of the bullet's flight. For this reason, the aim must be in front of the target; otherwise, the shot will fall behind the target.

Factors Affecting Lead

Factors that affect the amount of lead are the target's range, speed, and angle of movement.

Range. There is a time lag from the time a round is fired until the round strikes at the point of aim. This time of flight could allow a target to move out of the bullet's path if the round were fired directly at the target. Time of flight increases as range to the target increases. Therefore, the lead must be increased in proportion to the distance to the target.

Speed. A greater lead is required to hit a running man than a walking man because the running man moves a greater distance during the flight of the bullet.

Angle of Movement. The angle of target movement also affects the amount of lead required for target engagement. The angle of movement across the Marine's line of sight relative to the flight of the bullet determines the type (amount) of lead.

Types of Leads

Full Lead. The target moves straight across the Marine's line of sight with only one arm and half the body visible. This target requires a full lead because it moves the greatest distance across the Marine's line of sight during the bullet's flight.

Half Lead. The target moves obliquely across the Marine's line of sight (at about a 45-degree angle). One arm and over half of the back or chest are visible. This target requires half of a full lead because it moves half as far as a target moving directly across the Marine's line of sight during the bullet's flight.

No Lead. A target moving directly toward or away from the Marine presents a full view of both arms and the entire back or chest. No lead is required. This target is engaged in the same manner as a stationary target because it is not moving across the Marine's line of sight.

Point of Aim Technique

To engage a moving target, a Marine establishes a lead using a point of aim offset aiming technique. This technique uses predetermined points of aim to sector a man-sized target vertically, halfway between center mass and the leading edge of the target (both one point of aim and two points of aim) (see fig. 7-1 on page 7-8). The same units of measure can be applied off the target for holds of additional points of aim. To use the point of aim technique to establish a lead on a man-sized moving target at various ranges, speeds, and angles of movement, the following guidelines apply. See figure 7-2 on page 7-9.

- For a target moving at a distance of 15 yards away or less, no lead is required.
- For a target walking directly across the Marines line of sight (full lead) between 16 and 25 yards away, hold 1 point of aim in the direction the target is moving.
- For a target running directly across the Marine's line of sight (full lead) between 16 and 25 yards away, hold 2 points of aim in the direction the target is moving.

Methods of Engagement

Moving targets, although difficult, can be engaged by the tracking or ambush method or a combination of the two. See figure 7-2 on page 7-9.

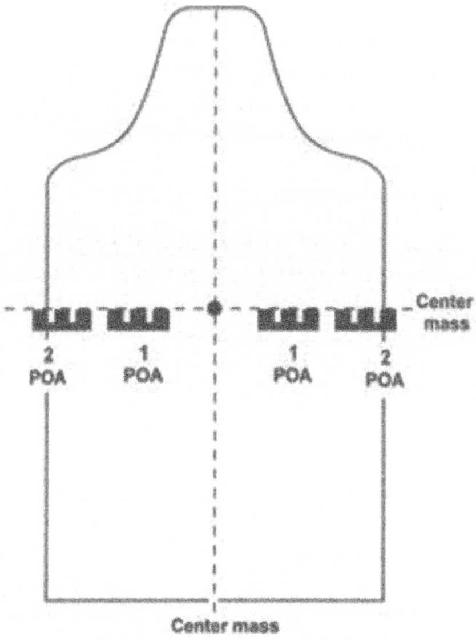

Figure 7-1. Points of Aim.

Tracking Method

In this method, the Marine "tracks" or follows the target with his front sight while maintaining sight alignment and a point of aim on or ahead of (leading) the target until the shot is fired. Sight picture is the aiming point in relation to the target while maintaining sight alignment (when a lead is established in moving target engagement, the sights are not entered on the target). To execute the tracking method—

- Thumbcock the pistol while presenting it to the target.
- Track the muzzle of the pistol through the target to the desired point of aim (lead). The point of aim may be on the target or some point in front of the target depending upon the target's range, speed, and angle of movement.
- Track and maintain focus on the front sight while applying trigger pressure and acquiring sight alignment.
- Continue tracking and applying trigger pressure while acquiring sight picture. When sight picture is established, engage the target while maintaining the proper point of aim (lead).
- Follow through so the lead is maintained as the bullet exits the muzzle. Continuing to track also enables a second shot to be fired on target.

Ambush Method

The ambush method is generally used to engage a stop and go target and when it is difficult to track the target with the pistol, such as in the prone position. With this method, the pistol is aimed at a predetermined engagement point ahead of the target and along its path, allowed to remain stationary, and fired when the target reaches the predetermined engagement point. The engagement point is based on the required point of aim (lead) to engage the target effectively. When the sights are settled, the target moves into the predetermined engagement point creating the desired sight picture. Once sight picture is established, the remaining pressure is applied on the trigger until the shot is fired. To execute the ambush method—

- Look for a pattern of exposure.
- Thumbcock the pistol while presenting it to a selected point of aim ahead of the target.
- Apply trigger pressure while obtaining sight alignment in the aiming area.
- Continue applying trigger pressure and hold sight alignment until the target moves into the predetermined engagement point and the desired sight picture is established.
- Engage the target once the sight picture is acquired.
- Follow through so the pistol sights are not disturbed as the bullet exits the muzzle.

Applying the Fundamentals of Marksmanship

The engagement of moving targets is a perishable skill that must be practiced frequently if it is to be maintained. The Marine must practice to develop the skill to calculate the point of aim (lead) and

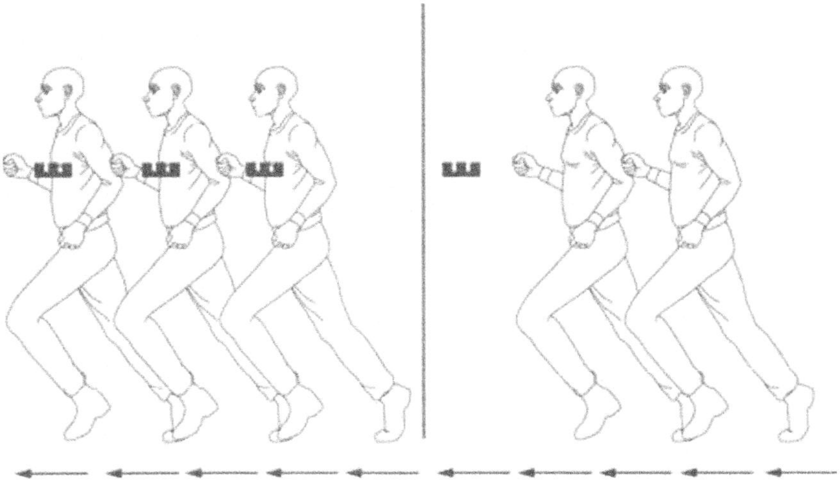

Figure 7-2. Moving Target Engagement Methods.

fire the shot while maintaining the proper point of aim (lead).

To engage moving targets using the tracking method, the Marine moves the pistol smoothly and steadily as the target moves. A stable position and firm grip are necessary to steady the pistol sights while tracking.

When using the tracking method, continue tracking the target while following through with the shot process so the point of aim (lead) is maintained as the bullet exits the muzzle. Continuing to track also enables a second shot to be fired on target if necessary. Concentration should be on continuing to track while applying the fundamentals. The fundamentals must be instinctively applied, allowing concentration on tracking the target and applying point of aim (lead).

Aiming

Sight alignment remains unchanged for accurate engagement of a moving target. The most common error when engaging moving targets is to focus on the target rather than focusing on the front sight. Sight picture is the point of aim in relation to the target while maintaining sight alignment. For both the ambush and tracking methods, sight picture is changed from the normal center mass picture, based on range, speed, and angle of movement of the target. It takes practice in moving target engagement to quickly establish the desired sight picture for a given point of aim (lead).

Breath Control

There is no difference in breath control when engaging moving targets; the breath is held to fire the shot.

Trigger Control

Trigger control is critical to firing shots while establishing and maintaining sight alignment and sight picture. Because the single action mode is the preferred method of engaging a moving target, the pistol is thumbcocked before trigger control is applied. The following guidelines apply:

- Do not use interrupted trigger control when engaging moving targets because the point of aim (lead) is lost or has to be adjusted if the trigger is held in order to re-assume the proper sight picture.
- Do not stop tracking as trigger control is applied. This causes the shot to impact behind the moving target. Trigger control should be uninterrupted while maintaining the point of aim so the trigger is pulled in one continuous, smooth motion to the rear.
- Do not interrupt trigger control when the target is masked from view. This is particularly true in the ambush method for targets that appear to stop and go when moving to and from cover. Trigger control should be uninterrupted.

CHAPTER 8
ONE-HANDED TECHNIQUES

In combat, a situation may arise where the Marine must engage a target by using only one hand. A Marine fires the pistol one-handed when—

- The Marine's other hand is occupied.
- A target presents itself so quickly and at a close range that the Marine does not have time to establish a two-handed grip and position.
- One of the Marine's hands is injured.

Note: The procedures in this chapter are written for right-handed Marines; left-handed Marines should reverse directions as needed.

Presentation and Shooting

Adjustment of the Standing Position

To fire accurately, the Marine must maintain the same stability and control of the pistol, manage recoil, and recover on target as effectively with one hand as with two. This is accomplished by adjusting the standing position as the pistol is presented to the target (see fig. 8-1).

The Marine's angle to the target determines what adjustments are made to the standing position. This adjustment is made while the pistol is being presented to the target by moving or sliding the feet to orient the body to the target. For stability and balance, the right foot may be slightly forward of the left foot. The further to the right of the Marine the target is the farther forward the right foot will be. To adjust the standing position when firing one-handed—

- Increase the muscular tension in the right arm and bend the elbow slightly to better manage the pistol.

Note: For some Marines, muscular tension in the firing arm and the position of the body in relation to the target causes them to cant the pistol inboard slightly. A slight inboard cant of the pistol and firing arm allows a natural bend in the elbow and allows a better management of recoil and recovery of the sights back on target. Any cant should be natural and not caused by excessive muscular tension. A slight cant does not affect shooting performance and should not alter the application of the fundamentals of marksmanship.

- Apply controlled muscular tension throughout the body to better manage the effects of recoil.
- Lean forward aggressively, roll the shoulders forward, and tuck the head into the shoulders.

Figure 8-1. One-Handed Presentation.

Note: The following step is done only during training to demonstrate how to stabilize the position and to simulate the Marine's other hand being occupied.

- Place the left hand in a fist on the center of the torso to increase stability. If the left arm is swinging freely it causes a corresponding movement in the pistol, reducing stability and control. Placing the left hand on the torso also ensures it is not covered by the muzzle of the pistol as the pistol is presented.

Searching and Assessing After Firing

Following one-handed target engagement, the Marine must quickly decide if it is to his advantage and if it is possible to free his occupied hand in order to place both hands on the pistol to search and assess. Whenever possible, the Marine places both hands on the pistol to increase stability and to be ready for re-engagement if necessary.

Reloading

Dry Reload

When only one hand is available, the dry reload can be performed as follows:

- Seek cover, if the situation permits.
- Press the magazine release button and allow the magazine to fall to the deck. If using the right hand, press the magazine release button with the thumb. If using the left hand, press the magazine release button with the trigger finger.
- Point the muzzle in a safe direction and position the pistol to facilitate loading the magazine. Depending on the firing position, this is accomplished using one of the following methods:

  Two-Knee Kneeling Position

  - Rotate the pistol so the magazine well faces up and the muzzle of the pistol is pointed away from the body.
  - Place the pistol between the thighs or knees and apply pressure to hold the pistol in place.
  - Withdraw a filled magazine from the ammunition pocket and insert it into the magazine well, seating it with the heel of the hand. See figure 8-2.

Figure 8-2. One-Handed Dry Reload: Two-Knee Kneeling Position.

Standing Position

- Rotate the pistol so the magazine well faces up and the muzzle of the pistol is pointed away from the body.
- Place the pistol between the thighs or knees and apply pressure to hold the pistol in place. Bend slightly at the knees to secure the pistol between the thighs.
- Withdraw a filled magazine from the ammunition pocket and insert it into the magazine well, seating it with the heel of the hand. See figure 8-3.

Kneeling Position (High, Medium, Low)

- Rotate the pistol so the magazine well faces outboard and the muzzle is pointed away from the body.
- Place the rear portion of the slide in the bend of the right knee and apply pressure with the leg to hold the pistol in place. For additional control, the Marine may drop to a two-knee kneeling position and place the pistol between his knees.
- Withdraw a filled magazine from the ammunition pocket and insert it into the magazine well, seating it with the heel of the hand. See figure 8-4 on page 8-4.

Prone Position

- The Marine places the pistol on the deck in front of him and against his body with the muzzle pointed in a safe direction. Alternately, the Marine may roll on his side and place the pistol between his knees, with the magazine well facing out.
- Withdraw a filled magazine from the ammunition pocket and insert it into the magazine well, seating it with the heel of the hand. See figure 8-5 on page 8-4.

- Grasp the pistol grip and remove the pistol from its secured location or from the deck.
- Press the slide release to allow the slide to move forward and chamber a round. If using the right hand, press the slide release with the thumb. If using the left hand, press the slide release with the index finger.

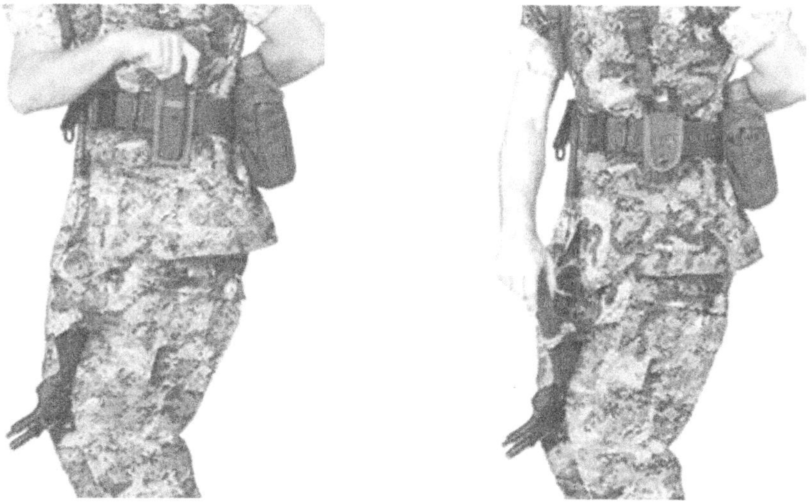

Figure 8-3. One-Handed Dry Reload: Standing Position.

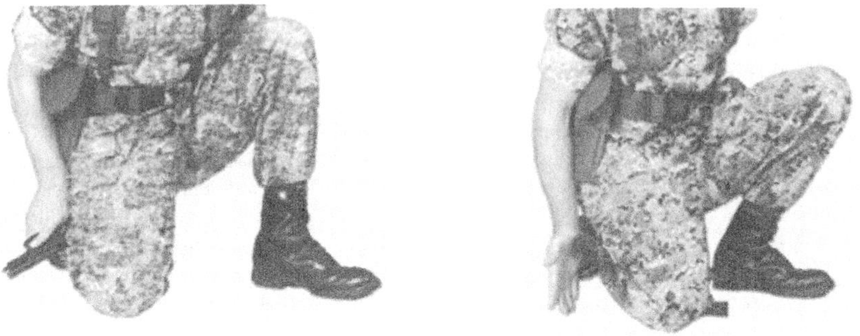

Figure 8-4. One-Handed Dry Reload: Kneeling Position.

Condition 1 Reload

When only one hand is available, a Condition 1 reload can be performed as follows:

- Point the muzzle in a safe direction and position the pistol to allow for removal of the magazine (e.g., between the thighs, in the bend of the knee, in the holster, tucked in the cartridge belt, on the deck against the body). Any position is correct if it allows the magazine well to be exposed.
- Withdraw a filled magazine from the ammunition pocket.
- Press the magazine release button and remove the magazine from the pistol. If using the right hand, press the magazine release button with the thumb. If using the left hand, press the magazine release button with the index finger.
- Insert the filled magazine into the magazine well, seating it with the heel of the hand.
- Stow the partially-filled magazine (e.g., inside the cartridge belt, in the cargo pocket).
- Grasp the pistol grip and remove the pistol from its secured location, keeping the trigger finger straight and off the trigger until ready to fire.

Remedial Action

Remedial action requires investigating the cause of the stoppage, clearing the stoppage, and returning the pistol to operation. When performing remedial action, seek cover if the tactical situation permits. Once a pistol ceases to fire, the Marine must visually or physically observe the pistol to identify the problem before it can be

Figure 8-5. One-Handed Dry Reload: Prone Position.

cleared. The steps taken to clear the pistol are based on what is observed:

- Remove the finger from the trigger and place it straight along the receiver.
- Lock the slide to the rear. To pull and lock the slide to the rear, push up on the slide stop and maintain pressure with the right thumb. (Push up on the slide stop with the left index finger if remedial action is being performed with the weak hand.)
- Secure the rear sight on the top of the cartridge belt or any other surface that provides the resistance needed to stabilize the pistol. See figure 8-6. While applying pressure on the pistol to keep the rear sight secured, push downward on the pistol in one continuous motion to lock the slide to the rear.

  Note: Ensure the pistol does not move to safe when locking the slide to the rear.

- Place the pistol in a position to observe the chamber.
- Correct the stoppage.

If there is a round in the magazine but not in the chamber, release the slide and observe a round being chambered. If you do not observe a round being chambered, tap, rack, and bang to perform remedial action:

<u>Tap</u>

- Keep muzzle pointed in a safe direction and place trigger finger straight along the receiver.
- Strike the bottom of the magazine against a hard surface to ensure it is seated. In the kneeling position, strike the magazine against the thigh (see fig. 8-7 on page 8-6). In the standing position, bring the right knee up while striking magazine against the right thigh (see fig. 8-8 on page 8-6). In the prone position, strike the magazine against the deck.

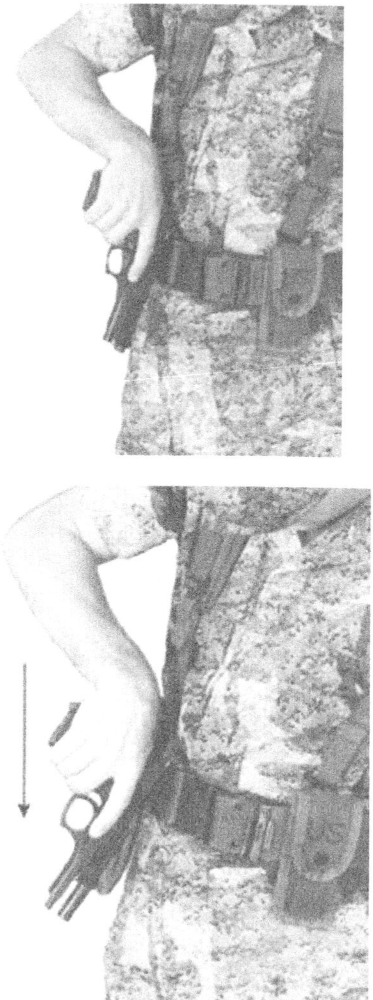

Figure 8-6. One-Handed Remedial Action.

Figure 8-7. Tap the Magazine Against the Knee: Kneeling Position.

Figure 8-8. Tap the Magazine Against the Knee: Standing Position.

Rack

- Rotate the pistol so the rear sight can be hooked on the top of the cartridge belt or any other surface (e.g., the edge of a table, wall, heel of the boot) that provides the resistance needed to rack the slide in a smooth, uninterrupted motion. Do not let the muzzle of the pistol cover the body. See figures 8-9 and 8-10.

Figure 8-9. Rack Slide Against Cartridge Belt.

Figure 8-10. Rack Slide Against Heel of Boot.

- Apply pressure on the pistol to keep the rear sight secured while pushing downward on the pistol to move the slide fully to the rear.
- Release pressure on the pistol to chamber the round.

Bang

- Recover the pistol on target, re-establish sight alignment/sight picture, and attempt to fire.

Presentation From the Holster With the Weak Hand

During combat, the Marine must be prepared to engage targets at any time. Therefore, the Marine must have the ability to present the pistol from the holster with his weak (left) hand if circumstances prevent him from using his strong (right) hand. He must apply the same smooth, controlled movement to present the pistol to the target. There are three methods for withdrawing the pistol from the holster with the weak hand. The Marine should practice each method to determine which works best for him.

Method One: Pistol Rotation

- Unfasten and release the D-ring with the left hand. See figure 8-11.

Figure 8-11. Method One: Release D-Ring.

- Wrap the fingers and thumb around the pistol grip so the index finger and thumb are around the base of the pistol grip. See figure 8-12.

Figure 8-12. Method One: Grasp the Pistol.

- Draw the pistol straight up until trigger guard clears the top of the holster. See figure 8-13.

Figure 8-13. Method One: Clearing Trigger Guard.

- Rotate the pistol so the magazine well faces the target and the trigger guard rests on the top of the holster. Ensure the trigger guard rests securely against the top of the holster. See figure 8-14 on page 8-8.

Figure 8-14. Method One: Rotating the Pistol.

- Maintain downward pressure on the pistol and slide the left hand around the pistol grip to establish a firing grip. See figure 8-15.

Figure 8-15. Method One: Establishing the Firing Grip.

- Grasp the pistol and continue to withdraw the pistol until the muzzle clears the holster and rotate the muzzle toward the target.
- Sweep the safety off with the thumb of the left hand while moving the pistol out toward the target. At the same time, place the trigger finger on the trigger and establish sight alignment and sight picture within the aiming area.
- Continue trigger pressure until the shot is fired.

Method Two: Hand Rotation

- Unfasten and release the D-ring with the left hand.
- Push the holster flap up and out of the way with the back of the left hand.
- Push the right hip out to facilitate grasping the pistol grip. Grasp the pistol grip between the holster flap and the pistol grip, the thumb is wrapped around the other side of the pistol grip against the backstrap. See figure 8-16.

Figure 8-16. Method Two: Grasping the Pistol.

Note: The cartridge belt may be pulled with the left hand to bring the holster closer to the body's center to facilitate grasping the pistol.

- Draw the pistol straight up while rotating the pistol so the magazine well faces inboard. Continue to remove the pistol until the trigger guard rests on the holster. See figure 8-17.
- Establish a firing grip and rotate the thumb to a position to operate the safety. See figure 8-18.
- Continue to withdraw the pistol until the muzzle clears the holster and rotate the muzzle to the target. See figure 8-19.
- Sweep the safety with the thumb of the left hand while moving the pistol out toward the

Figure 8-17. Method Two:
Trigger Guard Rests on Holster.

Figure 8-18. Method Two:
Establish Firing Grip.

target. At the same time, place the trigger finger on the trigger and establish sight alignment and sight picture within the aiming area.

- Continue trigger pressure until the shot is fired.

Method Three: Knee Placement

This method is particularly effective when firing from the two-knee kneeling position and the standing position.

Figure 8-19. Method Two:
Rotating the Muzzle.

- Unfasten and release the D-ring with the left hand.
- Wrap the fingers and thumb around the pistol grip so the index finger and thumb are around the base of the pistol grip.
- Draw the pistol straight up until the muzzle clears the holster and rotate the muzzle forward. The magazine well should face upward.
- Place the pistol between the thighs, applying enough tension to keep the pistol in place. The Marine must ensure that he does not disengage the safety while performing this motion and that the muzzle of the pistol is clear of the body. In the standing position, the Marine bends slightly at the knees to facilitate securing the pistol between the thighs.
- Remove the left hand from the pistol grip and rotate the hand around the pistol grip to facilitate establishing a firing grip.
- Grasp the pistol grip and remove the pistol from the thighs, rotating the pistol until the muzzle is pointed toward the target.
- Sweep the safety with the thumb of the left hand while starting to punch the pistol out toward the target.
- Establish a firing grip and continue to punch the pistol out toward the target. At the same time, place the trigger finger on the trigger and

establish sight alignment and sight picture within the aiming area.

- Continue trigger pressure until the shot is fired.

Transferring the Pistol From One Hand to the Other

Situations may arise in combat where the Marine must transfer the pistol to the left hand to engage a target because the right hand or arm is injured. To safely transfer the pistol from one hand to the other, perform the following steps:

Note: This procedure is written to transfer the pistol from the right hand to the left hand.

- Point the muzzle in the direction of the target and place the trigger finger straight along the receiver. See figure 8-20.

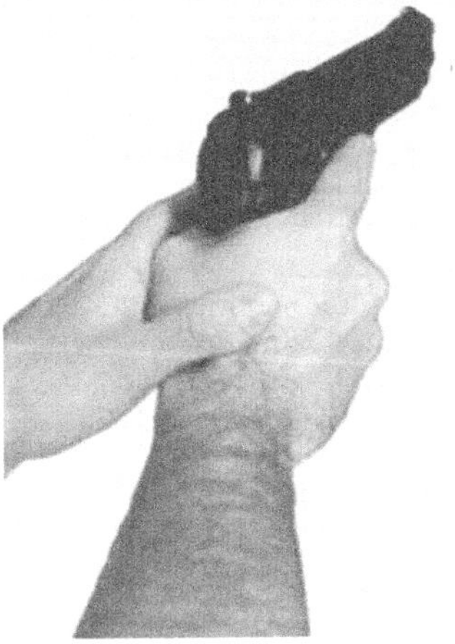

Figure 8-20. Transferring the Pistol—Step 1.

- Place the web of the left index finger and thumb under the backstrap of the pistol. See figure 8-21.
- Wrap the fingers of the left hand around the pistol grip while releasing the right hand grip.

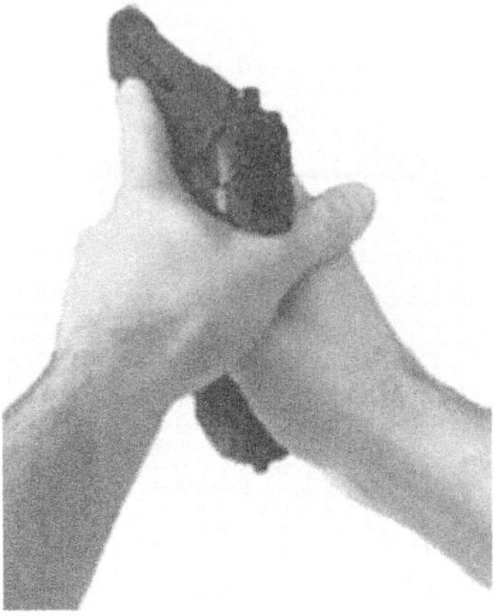

Figure 8-21. Transferring the Pistol—Step 2.

- Establish a firing grip with the left hand on the pistol. See figure 8-22.

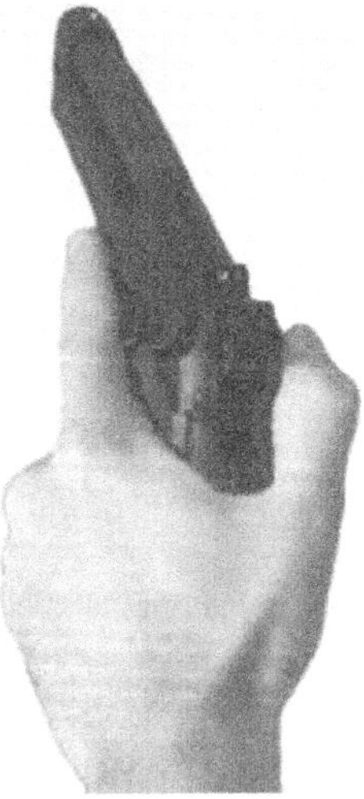

Figure 8-22. Transferring the Pistol—Step 3.

Chapter 9
Advanced Techniques

Some tactical situations are quite complex and require Marines to apply advanced pistol engagement techniques in order to effectively engage the enemy and survive. The ability of the Marine to rapidly shoot on the move and to turn and fire will be critical to a successful engagement.

Note: The procedures in this chapter are written for right-handed Marines; left-handed Marines should reverse directions as needed.

Shooting on the Move

When moving from one area of cover to another, the Marine may engage a target that presents an immediate threat. In this situation, the Marine relies on accurate fire to serve as his cover.

Moving With the Pistol

The likelihood of encountering a threat dictates how the pistol is carried while the Marine is moving. But regardless of the method of carry, the pistol's muzzle is pointed in the same direction that the head and eyes are looking—the eyes and muzzle move as one.

When carrying the pistol in the Alert position, the pistol is tucked in close to the body, finger off the trigger. This position allows the Marine freedom of movement and still allows for a quick presentation of the pistol.

If there is a high probability of encountering a threat, the Marine carries the pistol in the Ready position while moving.

Moving: The Glide Technique

To shoot accurately while moving, the Marine cannot run or walk fast. A normal running movement produces too much bounce in the Marine's body and makes it extremely difficult to use the sights of the pistol or to achieve the needed level of stability for accurate shooting. Therefore, the proper movement technique is similar to a glide and allows for accurate target engagement:

- A lower center of gravity is achieved by keeping the knees bent and the upper body erect.
- The bent knees also serve to absorb the shock generated by movement.
- The feet are not lifted as high as in a normal walk or run, allowing obstructions to be kicked out of the way.
- The feet and knees point in the direction of travel. If there is a need to engage a target off to the side, the Marine rotates at the waist to move the upper body in the desired direction. The upper body shooting position does not change.

The glide is not a technique that can be used for a long period of time or for long distances. Rather, the Marine can move in a normal manner until it becomes necessary to engage a target, he then slows and assumes the glide.

Engaging Targets: Using the Pistol Sights

When the Marine fires the pistol, he must concentrate on the pistol's front sight. The adherence to the fundamentals of marksmanship becomes even more vital when both the Marine and the enemy may be moving.

Continuing to Move

A Marine continues to move no matter what happens. Continuing to move makes it difficult for the threat to engage the Marine and can be one of the Marine's best defensive assets. When moving to cover or to a different position, the Marine should not stop to engage a target.

Reloading and Stoppages

If a stoppage occurs or a reload is required while moving, the Marine's primary consideration is to keep moving and seek cover. Ideally, reloads and clearing stoppages are performed behind cover.

Turn and Fire

The techniques for turn and fire allow a Marine to engage a target that is not directly in front of him. The key to turn and fire is smoothness and quickness of pivoting and presentation of the pistol to engage the threat. Turn and fire is based on the principle that wherever the head goes, the body follows.

Engaging Targets 90 Degrees to the Right and Left

The Marine turns the head toward the threat and identifies the target. Once the target is identified, the Marine maintains focus on the target for the rest of the presentation.

If engaging a target 90 degrees to the right, the Marine turns toward the target by raising the left foot while pivoting on the ball of the right foot. The Marine plants the left foot once he is facing the target squarely. The stance is natural. See figure 9-1. The Marine then engages the target.

Figure 9-1. Engaging Targets 90 Degrees to the Right.

If engaging a target 90 degrees to the left, the Marine turns toward the target by raising the right foot while pivoting on the ball of the left foot. The Marine plants the right foot once he is facing the target squarely. The stance is natural. See figure 9-2. The Marine then engages the target.

Engaging Targets 180 Degrees to the Rear

The Marine turns the head toward the target, looking over either the right or left shoulder, and identifies the target. Once the target is identified, the Marine maintains focus on the target for the rest of the presentation. See figures 9-3 on page 9-4.

If the Marine pivots to the right to engage the target, pick up the left foot, move it across the right foot, while pivoting 180 degrees on the ball of the right foot, plant the left foot, and squarely face the target in a natural stance. See figure 9-4 on page 9-4.

If the Marine pivots to the left to engage the target, pick up the right foot, move it across the left foot, while pivoting 180 degrees on the ball of the left foot, plant the left foot, and squarely face the target in a natural shooting stance. See figure 9-5 on page 9-4.

The Marine then engages the target.

Figure 9-2. Engaging Targets 90 Degrees to the Left.

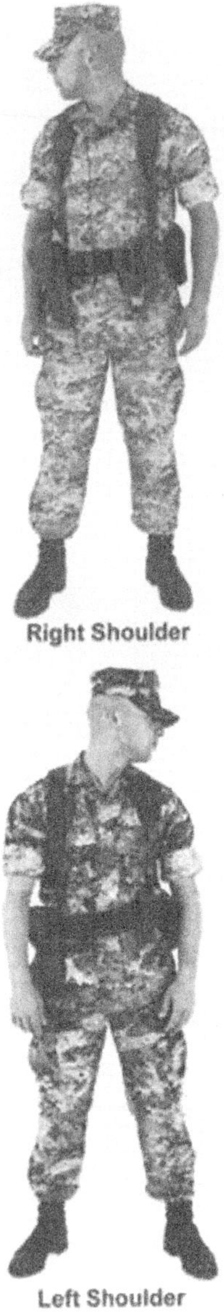

Figure 9-3. Engaging Targets 180 Degrees to the Rear.

Figure 9-4. Engaging Targets 180 Degrees to the Rear (Right Shoulder).

Figure 9-5. Engaging Targets 180 Degrees to the Rear (Left Shoulder).

Chapter 10
Low Light and Darkness Techniques

An effective combat marksman must be prepared to detect and engage targets under a variety of conditions. Factors such as terrain and opportunity often dictate that the Marine engage combat targets at night or under low-light conditions.

The fundamentals of marksmanship are employed for engagement of targets in low light and darkness just as they are in daylight. However, the principles of night vision and target detection when engaging targets at night must be applied.

Combat Mindset

In the stress of the combat environment, Marines must eliminate any hesitation, fear, or uncertainty of action and focus on the actions to fire well-aimed shots. This is important during low light and darkness when attention is more easily diverted because the sense of vision is reduced. This may create a shock of awareness as the Marine relies more heavily on his other senses. Every noise, movement, and muzzle blast appears intensified at night. The physical acts of shooting must become second nature to the Marine, so his focus will not be diverted from firing well-aimed shots. This is accomplished through mental preparedness and training until shooting actions become instinctive.

Night Vision Adaptation and Maintenance

Night Vision Adaptation

There are two methods for acquiring night vision.

The first method is to remain in an area of darkness for about 30 minutes. This area can be indoors or outdoors. The major disadvantage of this approach is that a Marine is not able to perform any tasks while acquiring night vision in total darkness.

The second method is to remain in a darkened area under a low intensity red light (similar to the light in a photographer's darkroom) for about 20 minutes, followed by about 10 minutes in darkness without the red light. This method produces almost complete night vision adaptation while permitting the performance of some tasks during the adjustment period.

Night Vision Maintenance

Because the eyes take a long time to adjust to darkness, it is important to protect night vision once it is acquired. To maintain night vision, the Marine avoids looking at any bright light and shields the eyes from parachute flares, spotlights, or headlights. When using a flashlight to read a map or other material—

- Put one hand over the glass to limit the area illuminated and the intensity of the light.
- Keep one eye shut to reduce the amount of night vision lost.
- Cover the light with a red filter to help reduce the loss of night vision.

Target Detection Techniques

Once the Marine has acquired night vision, he is prepared to locate targets. Some of the daylight observation techniques, such as searching for target indicators, also apply at night or in low light. But night observation techniques must allow for the limitations of night vision and the need to protect it.

Off-Center Vision

Because of the placement of the cones in the center of the retina and the rods around the edges, the angle at which the Marine observes an object at night affects how well it is seen. Off-center vision is the technique of keeping the attention focused on an object without looking directly at it. To maximize the use of the eye's rods that provide night vision—

- Never look directly at the object being observed.
- Look slightly to the left, right, above, or below the object.

Each Marine must experiment and practice this technique to find the best off-center angle. Typically, the best off-center angle is 6 to 10 degrees away from the object or about a fist's width at arm's length. See figure 10-1.

Scanning/Figure Eight Scan

Scanning is the use of off-center vision to observe an area or object and involves slowly moving the eyes in a series of separate movements across the objective area. See figure 10-2. To execute a figure eight scan, move the eyes in a figure eight pattern in short, abrupt, irregular movements over and around the area. Once a target indicator has been detected, focus should be concentrated in that area, but not directly at it. It is more effective to scan from a prone position or a position closer to the ground than the object being observed, creating a silhouetted view of the object. When scanning an area, look and listen for the same target indicators as in daylight: movement, sound (which seems louder at night than during daylight), and improper camouflage. Also, objects in bright moonlight/starlight cast shadows just as in sunlight.

Effects of Illumination

Both ambient (natural) light and artificial illumination can affect a Marine's perception of the target's distance and size and his night vision. The following situations produce less of an effect at the close ranges that pistol engagements occur, but they still affect target detection and engagement:

- Light behind or between the Marine and the target illuminates the front of the target and makes it appear closer than it actually is.
- Light beyond the target displays the target in silhouette and makes it appear farther away than it actually is.
- The introduction of light requires the eyes to make a sudden, drastic adjustment to the amount of light received, causing a temporary blinding effect that occurs when night vision is interrupted abruptly. Ambient light can also have the

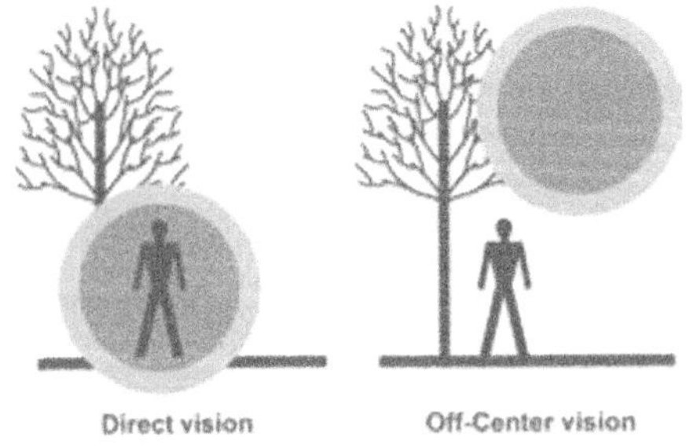

Figure 10-1. Off-Center Vision.

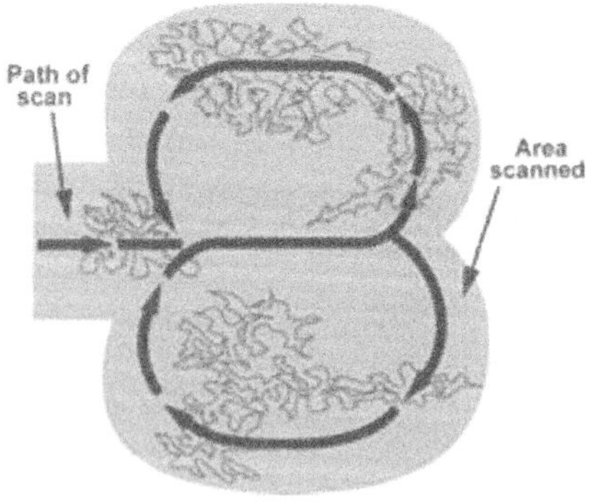

Figure 10-2. Figure Eight Scan.

same blinding effect; e.g., when a bright moon suddenly appears from behind clouds.

Acquiring Targets at Night

The Marine must keep both eyes open to get maximum visual coverage of the target area. Keeping both eyes open also improves depth perception and peripheral vision.

Sight Alignment/Sight Picture

Sight alignment and sight picture are obtained the same way in darkness as in daylight. There is normally enough ambient light to perceive objects as far away as 50 meters, especially if they are moving. However, when pistol sights are placed on a dark background, such as a camouflaged target, the Marine may not be able to acquire and align the sights clearly. Instead, the Marine may have to rely almost entirely on his presentation, firing position, and grip to get the pistol on target. To check for sight alignment or acquire the sights:

- Move the pistol just off the edge of the target to an area that provides a good contrast.
- Acquire sight alignment.
- Bring the sights back on line with the target while applying pressure to the trigger so the shot breaks once the sights are on target.

Flashlights

A flashlight can be used in low light and darkness to identify and illuminate targets. It is also used to acquire sight picture.

Types of Flashlights

There are a variety of flashlights used throughout the Marine Corps. They come in many different shapes and sizes; however, they are either straight/tubular or L-shaped.

Straight/Tubular

The straight/tubular flashlight is the most common. The body of the flashlight is a straight/tubular shape with the lens at the head of the flashlight. The on/off mechanism is located along the body or at the base of the flashlight.

L-Shaped

The L-shaped flashlight has a straight/tubular body, with the head/lens of the flashlight positioned at a 90-degree angle to the body of the flashlight. The on/off mechanism is located along the body or at the base of the flashlight.

Activation Devices

Depending on the type of flashlight, there are several ways that it can be activated.

On/Off Button

On/off buttons are located either along the body of the flashlight or at the base. On/off buttons are operated in a number of ways, depending on the flashlight.

On/Off Switch

On/off switches are mounted on the body of the flashlight. This switch has three settings that allow the flashlight to be either on, off, or in manual mode. In the manual mode, the flashlight is turned on by pressing and holding the button located next to the on/off switch.

Rotating Head

Rotating the head of the flashlight, either clockwise or counterclockwise, turns the flashlight on. Continuing this motion allows the width and intensity of the beam to be adjusted.

Types of Lenses

Clear Lenses

Clear lenses are used primarily for target detection and to illuminate the pistol sights. They are the most commonly used lens.

Colored Lenses

Colored lenses include red, amber, and blue. These lenses are used primarily for map reading and signaling, but can be used to illuminate the pistol's sights.

Target Detection

When a target is illuminated, the front sight may become silhouetted against the target, providing the sight picture needed to engage the target. During target detection, the Marine's focus should be twofold: to scan the area to identify possible targets and to assess the area to formulate a plan for engagement or cover (e.g., identifying the quickest route to cover, determining the best method for engagement based on terrain).

Searches

The Marine executes a hasty search immediately upon entering a new area in order to identify a threat that poses immediate danger. To execute a hasty search—

- Quickly scan the area with the flashlight taking note of obvious points throughout the area that could cover or conceal the enemy. Wherever the eyes move, the flashlight should move.
- Aim the flashlight beam on the ground about 8 to 10 feet in front of the Marine's location, this allows the eyes to follow the beam and quickly establishes a reference point for the light. Aiming the beam at the outer edge of the search area strains the eyes to find the beam and decreases the field of view.

The Marine executes a detailed search after the hasty search. A detailed search focuses on target indicators identified during the hasty search. To conduct the detailed search—

- Focus the most direct or intense portion of the flashlight beam on the target indicators.
- Move the beam of light slowly across the target indicator from right to left or left to right. Wherever the eyes move, the flashlight should move.

Considerations

During night operations, the Marine should use a flashlight whenever possible. A flashlight not only helps the Marine locate targets at night, but the beam can be aimed directly at the target once it is detected and temporarily blinds the target. This gives the Marine the advantage to react before the target does. However, the Marine must consider the following:

- Anytime a flashlight is on, the location may be revealed to the enemy. The Marine should keep the flashlight pointed out in front of his body to help avoid illuminating himself and revealing his position.
- Light shined directly from in front of the Marine at the target obscures the Marine.
- Light shined from the side of the Marine at the target illuminates the Marine.
- Light can bounce off surfaces and reflect back onto larger areas that the light is not focused on. Therefore, whenever a Marine shines a flashlight onto a surface, some of the light may reflect back and illuminate the Marine.
- When positioned at the outside corner of a room, building, or other cover, the Marine should avoid pointing the flashlight beam directly at the corner because the beam reflects off the corner and illuminates the Marine. To prevent this, the head of the flashlight should extend just beyond the corner.
- Many flashlights have adjustable beams that aid in target detection. The beam's intensity must be adjusted to provide the best illumination of the area depending on the distance between the Marine and the area of observation.
- Diffused light from a wide beam creates a softer light and illuminates a greater area, but the beam does not travel a great distance. A wide beam is best for observing larger areas at close range.
- Concentrated light from a narrow beam illuminates a smaller field of view, but the beam travels a greater distance. A concentrated beam is effective for observing a specific area or an area that is further away. Concentrated light

blinds the enemy and prevents him from focusing on the Marine or determining his location.

- When searching an area, the Marine can alter the position where he is holding his flashlight so that the beam is perceived as coming from various locations. For example, the Marine can turn the flashlight on from a standing position, quickly search for targets, turn the flashlight off, assume a kneeling position, and search again. This keeps the enemy from obtaining an exact location of the Marine's position.
- Keep both eyes open to get maximum visual coverage of the target area. Keeping both eyes open also improves depth perception and peripheral vision.

Target Engagement

Once a target is detected using a flashlight, the flashlight can be used to acquire sight picture and facilitate engagement of the target. The fundamentals of marksmanship are employed for engagement of targets in darkness just as they are in daylight. However, the Marine must apply the principles of night vision and target detection and must be able to employ a flashlight properly when engaging targets at night.

Grip

To engage a target accurately while using a flashlight, the flashlight must be held to maintain the pistol's stability, control, and recoil during firing. Normally, the left hand (two-handed grip) provides the pistol's stability and ability to manage recoil. However, when firing with a flashlight, the left hand holds and operates the flashlight. Therefore, some stability, control, and management of recoil is lost.

Securing the Flashlight

Most flashlights come equipped with a retaining loop located at the base for ease of transportation. To position the cord properly:

- Feed one end of the cord through the retaining loop of the flashlight and tie the ends of the cord together.
- Slip the left hand through the cord loop. The loop should be just big enough so that, if the flashlight needs to be dropped, it can be retained on the wrist.
- Rotate the flashlight until the cord is twisted to the desired tension and length necessary to best stabilize the rear end of the flashlight in the hand.
- Grasp the flashlight with the left hand in a position that allows it to be operated easily.

If the flashlight has a retaining loop, attaching a cord to the loop helps support and stabilize the flashlight when the Marine must fire the pistol with the flashlight in his left hand. The cord also helps retain the flashlight.

Acquiring Sight Alignment/Sight Picture

When holding a flashlight on a target, the tendency is to look at the target rather than the sights. Sight alignment is still necessary for effective target engagement. Using a flashlight to illuminate a target allows the Marine to—

- Acquire sight picture by silhouetting the sights against the target.
    - The Marine establishes sight picture by focusing the sights in the soft, diffused light area of the target.
    - At close ranges, colored lenses produce enough light on the target to silhouette the sights; at long ranges, colored lenses will not provide enough light off the target to illuminate the pistol sights. The Marine must be able to see the target by some other means, (e.g., contrast, moonlight).
- Acquire sight picture by illuminating the sights.
    - A colored lens can be held directly over the sights to shine light on the top of the pistol to illuminate the sights. This allows sight alignment to be established without revealing the Marine to the enemy. This method can also be used with a clear lens to make a precision shot at long ranges but the Marine risks being illuminated by the light.
    - A colored lens can be shined from directly behind the pistol sights to illuminate them. This

allows sight alignment to be established, but reveals the Marine's position to the enemy.

Target Engagement Techniques

Two-Handed Grip Technique. This technique is typically used with a straight/tubular flashlight. The pistol and flashlight must be side by side and level so the Marine can engage a target without making adjustments to the pistol or the flashlight. This technique is also effective for firing multiple shots because the flashlight and pistol recoil as a unit.

Whenever possible, the flashlight is just in front of the muzzle of the pistol so that it does not illuminate the pistol. Placement of the flashlight alongside the pistol may need to be adjusted depending on the size of the flashlight or the location of the on/off mechanism.

This technique is most often used with an Isosceles position (see fig. 10-3):

- Grasp the flashlight with the left hand, lens pointing down range.
- Wrap the thumb and index finger of the left hand around the body of the flashlight, thumb resting on the on/off switch to allow easy operation of the on/off switch without disrupting the grip on the flashlight.
- Extend both arms toward the target and bring the flashlight alongside the pistol so the fingers of the left and right hands touch.
- Wrap the bottom three fingers of the left hand around the fingers of the right hand, incorporating the flashlight into a two-handed grip.
- Apply isometric pressure against both hands to aid in stabilizing the pistol.

This technique can also be incorporated into a Weaver position; the more the left hand wraps around the right, the more the body can be angled and the left elbow bent. Apply push-pull pressure on the grip to stabilize the pistol. See figure 10-4.

Cross-Hand Technique. The cross-hand technique is used primarily with a Weaver position and can be used with either a straight/tubular flashlight or an L-shaped flashlight (see fig. 10-5):

- Grasp the flashlight with the left hand, fingers wrapped around the top of the flashlight and thumb wrapped around the bottom.

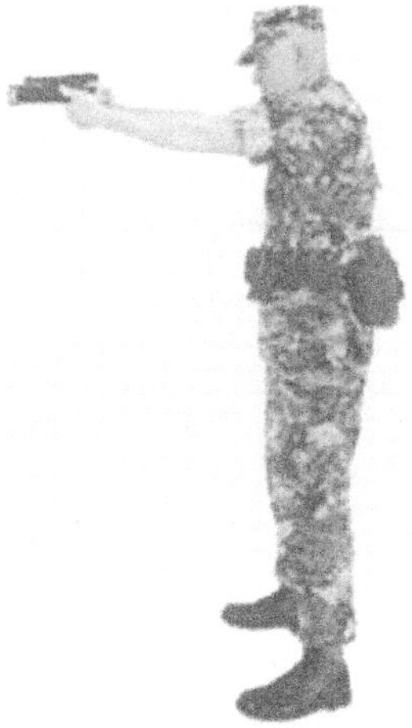

Side

Front

Figure 10-3. Two-Handed Grip Technique With Flashlight (Side and Front View).

Figure 10-4. Two-Handed Grip Technique With Flashlight (Weaver Position).

Figure 10-5. Flashlight Cross-Hand Technique.

- Place the thumb on the on/off button while maintaining a firm grip on the flashlight.
- Extend both arms toward the target and bring the left hand under the pistol so that the back of the left hand is resting firmly against the back of the right hand.
- Apply isometric pressure against both hands to aid in stabilizing the pistol.
- Extended the right arm fully and bend the left arm at the elbow.

Considerations for the Carry/Transport

When searching an area for targets, the Marine moves with the pistol at the carry or transport dictated by the threat level.

The Ready. If enemy contact is expected (contact imminent), the Marine carries the pistol at the Ready and carries the flashlight in his left hand and incorporates it into his firing grip. In the Ready, the Marine can search for targets and readily present the pistol and flashlight to the target for engagement without making adjustments to either the grip or the flashlight. See figure 10-6 on page 10-8.

The Alert. If enemy contact is likely, the Marine carries the pistol at the Alert and the flashlight in his left hand and incorporates it into his firing grip. The Marine must ensure the flashlight is not pointed at the deck because it illuminates the Marine. If necessary, the head of the flashlight may have to be tilted up to elevate the beam to increase the field of view. If the flashlight has been tilted, it has to be lowered so it is level with the pistol muzzle and the light shines directly on the target in order to establish sight picture. See figure 10-7 on page 10-8.

Holster Transport. If there is no immediate threat, the pistol is holstered and the Marine carries the flashlight with his left hand to search for targets. Should a target present itself, the Marine presents the pistol from the holster and engages the target. Time and distance to the target dictate whether the Marine incorporates the flashlight and his left hand into the firing grip.

Cross-Hand Two-Handed Grip

Figure 10-6. Flashlight Ready Carry (Cross-Hand and Two-Hand Grip).

Two-Handed Grip Cross-Hand

Figure 10-7. Flashlight Alert Carry (Two-Handed Grip and Cross-Hand).

Made in the USA
Middletown, DE
05 November 2026

36607798R00144